CH

D0713023

AN AMERICAN PROVENCE

AN AMERICAN
PROVENCE

Thomas P. Huber

UNIVERSITY PRESS OF COLORADO

Published by the University Press of Colorado
5589 Arapahoe Avenue, Suite 206C
Boulder, Colorado 80303

 The University Press of Colorado is a proud member of
the Association of American University Presses.

The University Press of Colorado is a cooperative publishing enterprise supported, in part, by Adams State College, Colorado State University, Fort Lewis College, Metropolitan State College of Denver, Regis University, University of Colorado, University of Northern Colorado, and Western State College of Colorado.

∞ The paper used in this publication meets the minimum requirements of the American National Standard for Information Sciences—Permanence of Paper for Printed Library Materials. ANSI Z39.48-1992

Library of Congress Cataloging-in-Publication Data

Huber, Thomas Patrick.
  An American Provence / Thomas P. Huber.
    p. cm.
  Includes bibliographical references and index.
  ISBN 978-1-60732-150-7 (hardcover : alk. paper) — ISBN 978-1-60732-151-4
(e-book : alk. paper) 1. Landscape assessment—France—Coulon River Region.
2. Landscape assessment—Colorado—Gunnison River Valley. 3. Landscape
assessment—Colorado—Paonia Reservoir. 4. Geographical perception—
France—Coulon River Region. 5. Geographical perception—Colorado—
Gunnison River Valley. 6. Geographical perception—Colorado—Paonia
Reservoir. 7. Coulon River Region (France)—Geography. 8. Paonia Reservoir
(Colo.)—Geography. 9. Gunnison River Valley (Colo.)—Geography. I. Title.
  GF91.F73H84 2011
  554.49'5—dc23
                        2011032202

Design by Daniel Pratt

20 19 18 17 16 15 14 13 12 11   10 9 8 7 6 5 4 3 2 1

All photographs were taken by the author.

It would seem appropriate to dedicate this book to the many people I have met in both the North Fork and the Coulon regions. I owe them much and can repay little. But I feel a stronger need to focus my gratitude on the person with whom this all started. My very, very French mother-in-law, Huguette Soichot Jones, raised my wife to love and honor France and Provence. I can no longer thank Huguette, but I can thank and dedicate this book to her daughter—my partner, best friend, guide, and translator.

CONTENTS

ACKNOWLEDGMENTS

I WANT TO THANK ALL the generous people of the North Fork Valley, especially Yvon Gros and Joanna Reckert. Also, John Cooley of the Valley Organic Growers Association was of tremendous help as I was preparing this book. The wonderful people of the Coulon Valley who were helpful are too numerous to mention, but I owe them and their beautiful landscape an immeasurable debt. I cannot even begin to thank my wife, Carole, for all her support, motivation, editing, and translation. Without you, this book would never have seen the light of day. Of course, any errors or misrepresentations are my own.

AN AMERICAN PROVENCE

PROLOGUE

W HEN I WENT OFF AFTER HIGH SCHOOL to get a college degree, it was pretty clear that my father wanted me to be a professional of some sort—a medical doctor would have been best, but a dentist (like my dad) or even an engineer would have been acceptable as long as I could make a good living and people would respect the profession I had chosen. I liked to solve problems and was pretty good at math, so I signed up as an engineering major my first semester at school. The school's curriculum was formulaic, to say the least. All the courses I took were part of the core except for those in my major and any overload "elective" courses I was willing to add on. In my second semester I signed up for one of the required core courses, called Introduction to Geography. I had always loved maps and dreamt about exotic places, but I had never given a thought to becoming a geographer—in fact, I had no idea anyone could be such a thing. Americans, unlike our European friends, do not have a tradition of geography as a basic discipline in universities and colleges. In Europe, geography is one of the most popular majors—this is especially true in England, Ireland, Germany, and France. But here in the United States, geographers are at best confused with geologists and at worst acknowledged with a perplexed look and a quick change of topic or, at cocktail parties, being asked to name the capitals of the fifty states.

Much to my amazement and my parents' deep chagrin, from my first moments in the class I was pleasantly stunned by my visceral connection with this topic. That was the end of my engineering career and the start of my much more exciting and often misunderstood career as a "professional" geographer. Of course there was much more schooling and degree taking and job hunting to undertake before I could be considered a professional, but in the end that is what I ended up being—a geographer.

My initial fascination with geography was hard to explain, even to myself. But as I learned more, I realized I was being shown the complex ways in which the world works and the intricate dance of causes and effects choreographed by place and space. The complex processes of the physical earth that set the stage for all life, including the human realm, were what I wanted to study and try to understand. I am still working on the understanding part.

This book is about that need to sense and understand this intriguing, quirky, and challenging world. As William Least

Heat-Moon writes in his book *Roads to Quoz*, "Human genesis arises from the land—from a place." The physical expression of "place" or of how we see a place can be summed up in the word *landscape*. In this book I will discuss two specific landscapes and, in a way, share my small gifts as a geographer with people who would like to understand a little more about land, place, and how we as humans live in and relate to landscape. Because of serendipity, discussed more in the first part of the book, I decided to look at two relatively small places that are seemingly from different worlds. But they both have the requisite human and natural elements that create a special place, and, to be honest, are two areas I have grown to love and wish to share.

One of these places is a small farming and ranching area along the valley of the North Fork of the Gunnison River in western Colorado. I first came to Colorado in the mid-1960s and adopted it as my home, or perhaps I was adopted by it—I am not sure I had a choice because we cannot really choose who or what we love. I have lived in, and worked extensively along, the Front Range and in the high mountains of the state for years, but only relatively recently have I spent large amounts of time in, and devoted study to, the state's far western regions. I discovered I was missing a lot. For example, the North Fork of the Gunnison in western Colorado has a physical environment and a human landscape that are at least as interesting as those of other areas I had been studying; in addition, the region is rapidly changing and developing into a cultural and economic magnet. The rest of the state may not see this transformation, but it is happening and it is happening quickly.

The second place is inspired by my wife's mother and her family. My mother-in-law was born and grew up in Marseille, France. All of her family members still live in southern France.

While visiting them, my wife, Carole, and I would travel through the almost mystical landscapes of Provence. Carole and I visited these places many times over a period of several years. Whenever I mention in the book that "we" did this or that, she was the one putting up with my geographical whims. These landscapes captured me much as Colorado did many decades ago. When I am in Hotchkiss or Paonia, I cannot help thinking about Provence; when in Provence I repay the compliment by using these places in western Colorado as my comparative landscape. Why is that? What are the factors that keep these images popping into my head? How can two sets of peoples separated by thousands of miles and different languages be so much alike and not know it? These questions are the basis for what follows.

I would like to think that everyone is as interested in the world's myriad landscapes as I am, but that is probably not the case. I am not writing this book for those who specialize in any particular discipline, even geography. I hope those who pick up this book and enjoy its contents will be people who are, most of all, curious—curious about a lot of things and eager to find out how different places became different and perhaps to learn that they are not so different if one looks at them with the right questions in mind. The people I would like to have read this book are the kind of people I would like to have long conversations with over dinner and maybe a little wine. This book is my surrogate invitation to you to a Provençal–North Fork landscape soirée.

Seeing, sensing, and understanding a landscape have common factors for everyone. The way the land lies and how the rock and soil create the foundation of a place, the patterns the vegetation takes as it responds to this foundation below and the climate above, and the way the land and vegetation mildly coerce humans into the ways they use that space are all inherent in one's sense of place. A *New York Times* article celebrating Wallace Stegner's 100th birthday put it well: "It's worth remembering another lesson of [Stegner's] life to choose authenticity over artifice. If you don't know where you are, he said . . . you don't know who you are." There is also a kind of phenomenological aspect for each of us in our response to a landscape. One person's bucolic scene is another's smelly, dirty farm. Each of us reacts to a landscape at least slightly differently. We all have our favorites; the ones in this book are some of mine that I wish to share with you.

Woven throughout this book is an underlying acknowledgment of the practical, yet natural, way many of the people in the two valleys live and work. I talk about organic farms and the food lovingly grown on them; I emphasize the role of nature in the ways viticulturalists decide where their vineyards will be placed on the earth. In a more academic work, I would probably write extensively about how these practices are all a part of "sustainability" and how it is inextricably linked to place—but the word *sustainability* is often overused and more often misunderstood, and I will keep the use of it to a minimum. So I simply report in the vernacular language the people of the two valleys use. Many inhabitants of both valleys are immersed in the organic farming movement, but they seldom talk about being leaders or in the forefront of a movement, even though they are the tacit leaders in their own communities. This connection to the land is simply how they view their place(s).

PLACES

---

*The voyage of discovery is not in seeking
new landscapes but in having new eyes.*
—MARCEL PROUST

*Colorado men are we, from the peaks granite, from the
great sierras and the plateaus, from the mine and from the
gully, from the hunting trail we come. Pioneers! O Pioneers!*
—WALT WHITMAN

ECAUSE I AM A GEOGRAPHER, I cannot stop looking at, thinking about, or visiting places. Perhaps some
kind of genetic disorder compels me to go to places,
to study places, to compare places. Other geographers
appear to share my malady, and they tend to use the word *place*
with a whole collection of meanings non-geographers might not
appreciate. We geographers see place as the interaction of all of
a location's physical characteristics, including soils, vegetation,
climate, and geology—much like an ecosystem except broader
and of a much larger scale. We also think about place as the
nexus of human occupation of and habitation on the land. In

this context, as characterized by *National Geographic* and others, place is "space endowed with human meaning." Place may even be mythical or spiritual or psychological. I, like my geographic colleagues, see place in all these ways in our attempt to make sense of the world.

Even though I understand place through a trained geographer's eyes, what follows is not a scholarly treatise on the subject. Rather, I offer in these pages a personal exploration of place, my attempt to endow meaning on two simultaneously diverse and yet, to my eyes, similar landscapes. Although this is not an academic study, I still see place and all its meanings and revelations as a geographer might—in this I am powerless to intervene.

The French use the word *milieu* to speak of place in the more inclusive way we geographers sometimes use. I like using *milieu* for at least two reasons. First, it is such a rich term in its complexity and its appropriateness as a geographic word. Second, because one of the places central to this book is in southern France, it just feels right to use this lyrical French term, which incorporates the physical setting of place with how people connect and sense a place.

The two places, or different milieux, in this book are perhaps what could be called *vernacular* or common landscapes and at the same time unique and special landscapes, depending on the viewer's mind-set. I have chosen to reflect on these two spots on Earth for idiosyncratic reasons. My professional and personal lives are intertwined in these places and their landscapes. These two seemingly disparate dots on the world map are similar in so many ways (and dissimilar in a few). Some of these ways are subtle, some not, but the personal and professional convergence drew me in with an intriguing intensity. It was almost as if I was urged on by some internal voice to look more intimately at these places. I literally put aside all of my scholarly projects and dove in with head and heart to see where this would lead.

I hope my passion for the land and the people will come through in the book. Perhaps my look at the North Fork and the Coulon will spur you, the reader, to look at your own personal milieux with a new eye. I would also hope that the book will spur you to travel to places that may be special to you because of your own curiosity and for your own reasons. Maybe instead of motivating you to travel, the book will motivate you to become more intensely interested in your place in the world. As Wendell Berry might say, to become more deeply local. In either case, place is important and is part of who we are. Books about places serve as surrogates for going to those places, but my advice is to use this book as a motivator to instill within yourself a more insightful sense of where you are.

This book was born on an early autumn morning a few years ago. Carole and I had gone to Hotchkiss on a whim. The United States was about to invade Iraq. Our government was unhappy with our French "allies" because they would not cooperate with US intentions. Americans started doing silly things like calling French fries "Freedom fries." One Sunday morning my wife and I read a story in the *Denver Post* that highlighted a particular inn and vineyard just outside Hotchkiss that were owned and operated by an Americanized Frenchman and his New York–born wife. Carole, whose mother had grown up in Marseille, is half French. We decided that this innkeeper could use our meager monetary and psychological support.

We stayed at the Leroux Creek Inn on that initial trip. I had just awakened the morning after our first night and walked into the inn's communal hallway. I looked southward out the large, second-floor window, and the valley of the North Fork of the

Gunnison River filled the landscape in front of me. The inn itself sits on Rogers Mesa with its mix of orchards, farm fields, pre-pubescent vineyards, and piñon-juniper woodlands; the large valley of the North Fork, which runs east-west, is down the hill to the south. Just to the north of the inn is the much bigger and higher Grand Mesa—purportedly the largest flattop mountain in the world. No one can prove or disprove this statement because there is no single definition of what a mountain is or how it is delineated, a point made in entertaining fashion in the film *The Englishman Who Went up a Hill and Came down a Mountain*. So the superlative remains as part of the milieu of this area of Colorado. Far to the south looms the plateau of Fruitland Mesa. Beyond this elevated landform, the main channel of the Gunnison River has carved the Black Canyon of the Gunnison (now home to Black Canyon of the Gunnison National Park), a deep incision in the earth that is visible only if one stands very near the edge. Otherwise the plateau looks benign, just another run-of-the-mill piñon and juniper–covered upland of the American Southwest.

At that moment in 2003, though, my thoughts or vague impressions were focused on the near distance, on the young vineyard lovingly established by the inn's owners, Yvon Gros and Joanna Reckert. The vineyard is small, barely four acres, but the rows of vines marched away from the inn in perfect precision, the stems heavy with fruit. For an instant my sleep-addled brain found itself in Provence—not the glitzy Côte d'Azur on the French south coast but the rural farmlands and small hilltop villages of the Vaucluse and the Luberon of northern Provence. An instant later the mental fog lifted and I was back in western Colorado but wondering why the Provençal image had not flashed into my mind sooner. My geographer's sense of place—

or places in this case—confused the two because they are so similar in an array of ways that were obvious once I really looked at the Hotchkiss landscape and the way people were using their small piece of the earth. That flash of sleepy insight set in motion my desire to chronicle these two complex, intricate, and intimate landscapes or places or milieux and share them with others through a geographer's eye and mind. What started out as a simple comparison of two like places in distant locations turned into a more complex and interesting task—a personal adventure, a phenomenological and experiential association with place. Much is similar between the two regions—the light, the valleys, the climate; and much is less so—the history, the geology, the physical makeup of villages. But the earth-bound feel of the land always comes through in both the North Fork and the Coulon.

Both of these places are political and cultural outliers. Neither place is central to the life of its society at large. Rural Provence might seem sophisticated to those from the McTowns of the United States, perhaps because we have been conditioned to view everything French as urbane. But to the French the Coulon area is not even a mere afterthought compared with Paris, Bordeaux, Lyon, or even the Côte d'Azur on the southern Provençal coast. It is a more blatant snub than a case of just being ignored. As an example, my mother-in-law would bristle whenever she was accused of having even the slightest Provençal accent. Similarly, Delta County, Colorado, is hardly a blip on the radar screen of the state's populous Front Range or in the halls of the legislature in Denver or for those whose image of Colorado is one of high mountains and skiing. Part of each place is wild, uncivilized, and in some cases brutal, at least during winter when the two seem far from the comforts of paved roads and cozy habitation. The highlands of the Vaucluse and the Luberon, as well as their counterparts, the Grand Mesa and the West Elks, can be as rugged and isolated as anyplace in France or the lower forty-eight states.

Many aspects of place affect us on a personal level; most of them are common characteristics geographers use to look at a region's land and people in an attempt to evoke the soul of a place. The first foundation of place, and the one I will start with, is the physical setting and how it influences what people do on the land. In subsequent chapters, villages, food, wine, special characteristics, and other components of place will be covered in turn. As a geographer and a traveler, I have found that my geographer's eye has made my journeys more interesting and arresting.

One of the most peculiar and special similarities between the two places discussed here is the quality of the light. This is especially true in the summer and fall when the sky is clear, when even the distant hills seem near enough to touch if only we reached out. The landscapes are writ large, and the open, hemispheric skies pull in those faraway elements to make expansive and intimate places at once. The quality of the light in Colorado comes from a combination of nearly arid air and high elevation. Very little light is scattered, so haze levels are low and the tones of the "blue" sky are a deep cyan color. This is especially true if we look to the northern sky in the Colorado high country—a searing, deep, unique cobalt blue is a sure sign we are somewhere above 6,000 feet and away from urban pollution.

In Provence, on the other hand, the humidity is higher and the elevations are lower, so there must be an alternative expla-

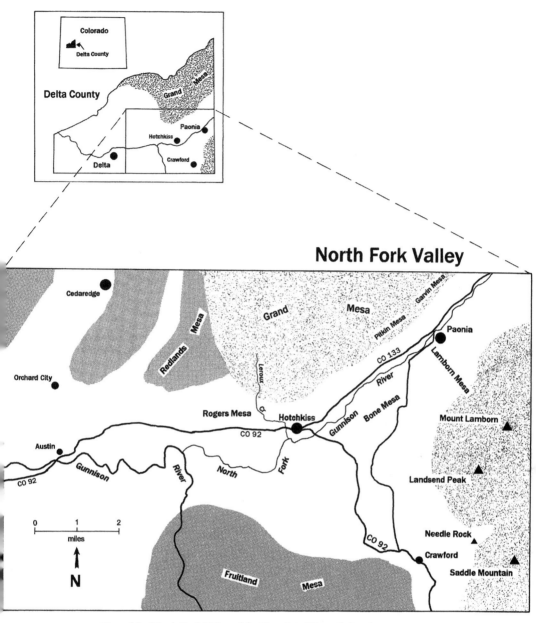

Map of the North Fork Valley of the Gunnison River, Colorado.

nation. The famous French impressionist painter Paul Cézanne saw the mix of intense sun and the vibrant patchwork of colors of the land that created intense and warm models for his art.

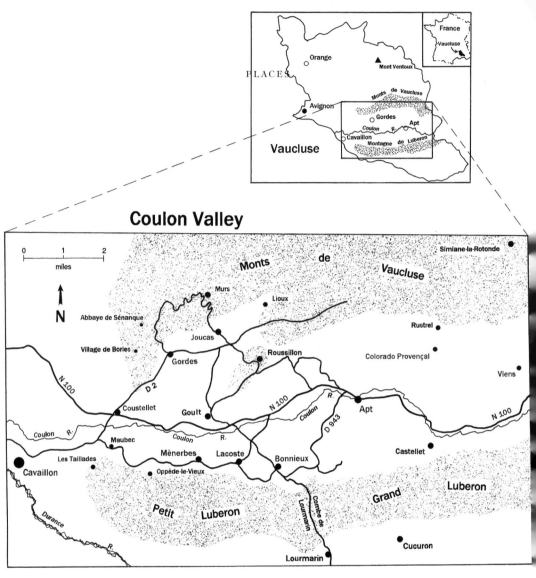

Map of the Coulon Valley in the Vaucluse, Provence, France.

In fact Provence's brilliant sun was too intense for Cézanne and his contemporaries during the heat of midday. They did most of their work in the early morning and evening light. Light was not the only reason Cézanne came to Provence, though. He purposely chose to move back to this rural, provincial part of

France. Author Nina Maria Athanassoglou-Kallmyer talks about Cézanne and many of his contemporaries in the late nineteenth century and their common move away from Paris: "[This self-exile] thus reenacted a practice common among contemporary artists forsaking the capital for one of the many preindustrial, unspoiled 'elsewheres.'" Cézanne was looking for what he and other impressionists felt was the authentic artistic milieu provided by the natural environment (especially the light), as well as wanting to retrieve the traditions and cultures lost in the cosmopolitan world of big cities.

The skies of Provence and Colorado do look remarkably alike when the cold, fresh, and strong mistral wind coming down the Rhône Valley blows the haze away or the strong up-valley/down-valley winds follow the course of the North Fork to clear the air. At those moments Provence has the sharp edges and clean lines found in western Colorado. These skies of brittle blue go a long way toward making these two places examples of the "elsewheres" Athanassoglou-Kallmyer describes.

Another reason I confused the two places on that autumn morning is that, in terms of the land as well as the sky, they look remarkably alike. For example, each valley is drained by a locally important river running from east to west. In the case of Hotchkiss and Paonia, it is the North Fork of the Gunnison River. Although this river is a mere "fork" of another, larger stream, it is a significant waterway for this semiarid western slope of Colorado. The North Fork rises high in the West Elk Mountains, where several modest creeks merge just south of Paonia Reservoir about twenty miles northeast of Hotchkiss. In spring the North Fork roars with water levels up to or above the channel banks. In the late fall, on the other hand, one can wade across the boulder-strewn bed without getting his or her

The North Fork of the Gunnison River as it flows past Paonia on the way to its junction with the main Gunnison River at Delta.

The smaller, but no less vivid, Coulon River from the Pont Julien at the eastern end of the Coulon Valley.

thighs wet. The North Fork merges with the main channel of the Gunnison River at Delta, about twenty miles west of Hotchkiss. The Gunnison itself joins with the famous Colorado River (formerly known as the Grand River) another thirty miles west, at Grand Junction.

The Provençal waterway that mimics this Colorado river has two names, neither of which I have been able to determine as official. East of the market town of Apt, it rises from the eastern end of the Monts de Vaucluse as le Calavan Rivière. After it flows through Apt, the local name changes to le Coulon Rivière. This might be a subtle indication of how the French (and rural Americans) view their landscapes as locally owned and not co-opted by the central government mapping agencies. One might not even notice the Coulon in midsummer because there is so little water in it. During my first crossing, I wondered what small stream this was and where the valley's major river flowed. As I drove farther across the valley bottom, I realized that this little stream was the Coulon. But as in all Mediterranean climate areas, although dry and sandy during the long, hot summer months, the riverbed fills rapidly in the wet winter and following snowmelt in early spring. Le Coulon empties into le Durance Rivière and then into the major river of southern France, the Rhône, just south of Avignon.

The valley floors of both places are vibrant agricultural areas, irrigated in the dry summer seasons by waters from the main rivers and side streams flowing from the higher elevations to the north and south or from wells drilled deep into ancient aquifers. Provence in particular has a mature farming culture that goes back hundreds, even thousands, of years. The landscape reflects this loving and intimate care over the centuries. The North Fork of the Gunnison is also a blossoming agricultural area but with a much shorter farming history that is barely 100 years old, first domesticated by the "Anglos" who settled there in the nineteenth century. The North Fork is a less intensively cropped land with large hay fields intermingled with small farms, orchards, and the more recent burgeoning vineyards. In nearby areas indigenous peoples have grown crops over the last 1,000 years or so, but no one has found much evidence of this ancient farming in the North Fork Valley today. The most famous of these ancient farmlands in western Colorado are now in Mesa Verde National Park, a little over 100 miles to the southwest. Although families have farmed the North Fork Valley for generations, I still see this landscape as a quickly adapting adolescent trying to figure out what it wants to do when it grows up. The Provençal landscape, however, changes just rapidly enough to remain interesting. There are places, for example, where old vines have withered and died from lack of attention, poor land for vineyards, or poor viticultural practices. Other fields have been planted only recently amid cherry trees or apricot groves. But mostly the vines are old, the trees mature, and the fields well tended.

Although the two areas have their share of bucolic landscapes, they are both working family farm regions, with all that entails. In each place I have seen ramshackle buildings, tractors clogging the narrow country roads, cars up on blocks, and rusting farm machinery in the yards—the latter provide storehouses of replacement parts for some future exigency.

When I am looking at either place, especially in the early morning or late evening light with the sun's slanting rays and long shadows, there are two things I cannot help but notice. Both places are lands of intricate complexity, and they both have the palpable character of land that has been in use for countless generations. In the case of the Coulon, human artifacts litter the land. In the North Fork, people have used the land for thousands of years but have left their human artifacts only during the last 200 years or so. They are both what I might call grounded palimpsests—places where cultures and history are multilayered and superimposed. This is especially true of Provence. The name *Provence* comes from the Latin word *provincia,* coined by Caesar as the Roman legions moved up the Rhône Valley from the city the Greeks established at the site of what is now Marseille. There are cultures layered upon other layers from the Greeks, Romans, Goths, Franks, Vandals, Saracens, the Provençal, and the modern French, to name the most well-known.

At times these layers are so subtle that they are hard to peel apart. As an example, the ruins of the infamous Château du Marquis de Sade are nearly seamless with the rest of the village of Lacoste. The stonework of the village matches the stonework of the chateau, and the village and the chateau share the steep slopes that lead up to the latter. Many of the currently occupied buildings in the town and the chateau were built at the same time, in the tenth century. The fields that now surround the hill of Lacoste have been tended for centuries. There is little in evidence here that shows that the Romans or Greeks were in the area, but only a few miles east sit the wonders of Roman architecture in Orange and Avignon.

The Golden Triangle of Hotchkiss, Paonia, and Crawford in the North Fork Valley has its own long history of human occupation and use, but that history has been less intense and its settlement much more sparse, so it differs from the intense nature of Coulon's human occupation. Little physical evidence exists concerning who occupied (or at least traveled through) this land prior to the last few centuries. The Ancestral Puebloans (Anasazi) populated the lands to the south and west in places such as Mesa Verde, Hovenweep, and deep into New Mexico. But there is little to no evidence that they came as far north as the North Fork Valley. There is a substantial oral history narrative by

This sign is a nearly forgotten reminder of the Dominguez-Escalante expedition that came through the North Fork Valley in 1776, long before Anglo settlement changed the land to the present bucolic setting.

Native Americans such as the Utes who have been in the valley for several centuries. The written record, however, only begins with the migration and exploration of Europeans who moved there in the late nineteenth century. A major exception was the Dominguez-Escalante expedition, which came through the area in 1776 on its ill-advised mission to find an easy land route from Santa Fe to California. As far as the specifics of settlement, Enos Hotchkiss and Sam Wade first brought fruit trees to the North Fork of the Gunnison in 1881 and 1882. This was the beginning of the land homesteading of the fertile valley of the North Fork and the orchard culture that still exists today.

Except for a few buildings that date back to the late nineteenth century, little in Hotchkiss's built environment is recognizable from its past. Generally, the remnants that do exist date from the early to mid-twentieth century, although things are changing fairly rapidly. Some of the changes in the town reflect a shifting economy, such as the old house that is now an art gallery. Others are indicators of the diverse ethnicity there and in the rest of the United States, such as the Middle Eastern restaurant on the main street. But what is most visible is the move to take the ubiquitous pasture of the valley bottom and create small, organic "farms" on two to three acres of land. After spending a considerable amount of time in the valley, I began to see these small vegetable and flower operations interspersed with orchards and newly planted vineyards. Even some of the lower-elevation piñon-juniper woodlands are being cleared for grapevines and organic vegetables.

One of my favorite things about the Hotchkiss area that is now visible and that has occurred for centuries in the Provence region is the complexity of the landscapes. Land use in each area looks like a patchwork quilt. There is a similarity in their diverse

A typical scene looking over large forest areas of piñon-juniper woodland in the uplands above the North Fork.

uses—both places have fertile valley soils developed from deposits of material that has eroded from the hills to the north and south. These soils are often deep and easily worked, and they produce verdant crops if water is available during the growing season. The uplands in both places are covered with drought-resistant plants. When I look at these hills from a distance, I can barely tell the difference between those of Provence and those above Hotchkiss and Paonia. During my years in Colorado, I have developed the knack of recognizing specific forest types from afar; but I cannot distinguish the forms, colors, and structure of these two landscapes. The piñon-juniper, Gambel oak,

The Vaucluse garrigue *has semiarid-adapted vegetation similar to the piñon-juniper woodland, including pines and junipers of various other species. The area's cliffs and mountains are even more rugged than the mesas and mountains above the North Fork.*

and serviceberry of the mesas above Hotchkiss are replaced in Provence by the Aleppo pine, green oak, and English holly; and the searing summer sun and cold winter winds are at least as intense in Provence as they are in Colorado.

But neither of these places is merely a bucolic setting for some quaint, nostalgic ideal. These are both places, milieux, of contrast, contradiction, and complexity. Although the valleys are often lush, at least where irrigation water is available, there is a much less civilized, even brutal beauty to the bordering uplands. High above the North Fork Valley, for example, rises the Grand Mesa, which was forged from the molten inferno of volcanic lava (basalt) flows. We can barely imagine the sight of this flowing rock unless we go to the island of Hawaii during one of its many eruptions. The lava is now a cold, frozen fossil of once liquid rock, and the slopes are now steep.

The Vaucluse parallels these rugged and intense landscapes. The rock of the uplands is limestone, not basalt, but the evidence of past geologic violence is still frozen in time here as well. All one needs to do is look at the Cliffs of Lioux (called the Falaise de la Madeleine by the locals), for example. These cliffs are over 2,000 feet (700 meters) long and 330 feet (100 meters) high and were abruptly created during a violent collision between two tectonic plates around 35 million years ago.

When one visits either the Vaucluse of southern France or the Golden Triangle of western Colorado, he or she cannot help but notice how warm it is during the parched summer months, with the intense sunshine and few shading clouds. One day in Apt after a leisurely outdoor lunch (sitting in the shade, of course), we returned to a car that had an inside temperature of 130° F. The only people who do not dart from shady spot to shady spot along the streets of a Provençal town are sweaty tourists (English most likely—at least according to Noël Coward) who do not have the sense to get out of the midday sun. Everyone eats their prodigious lunches outside on patios. But there is always a battle for the shady tables or at least for a seat near one of the ubiquitous misters at Provençal patio cafés that squirt diners with cooling water that evaporates quickly and lowers their spiraling body temperature.

Travel in winter in either of the two valleys can provide a broad spectrum of possible experiences. In both the North Fork and the Coulon, the cool, crisp winter days can be a refreshing and invigorating antithesis to the oven-like summers. Cool breezes, little snow, clear skies, and the promise of another tomorrow are typical. But the spectrum has another side as well. The North Fork can be inundated by heavy snow, strong winds, and icy roads for days on end. In the Coulon the nemesis that is always lurking in the winter is the mistral. The cold, hurricane-force winds roar down the Rhône into Provence from the high Alps. Provence is renowned in France for these ferocious, cold winds, which can lower temperatures by 30° F in fifteen minutes, last for days, and make life miserable for everyone and everything—including the grapevines. While driving through the vineyard areas in the Coulon Valley on one of my early trips to the area, I could not figure out why all the grapevines were only four or five feet tall. They looked like some kind of midget variety compared with vines I had seen in many other parts of

the world, including Hotchkiss. I subsequently learned that
they are pruned and trained especially so the mistral will not rip
them from the ground. Taller vines could not withstand days of
the 60-mile-per-hour (100-kilometer-per-hour) winds attained
by this usually unwelcome winter visitor.

Rain and snow in both places can be intense yet sporadic
and often leave impressions from which memories are forged.
While at another small town located along the Coulon Valley
floor, during one week in July we experienced a series of afternoon thunderstorms on consecutive days that blackened the sky,
cooled temperatures, and dropped a small amount of rain. For
six or seven days in a row, we could have set our watches by the
consistent timing of the storms. We were told by the locals that
this was very uncommon for a summer week. I tend to believe
them, although when people tell me the bad weather I am experiencing wherever I am is unusual, I rarely trust their objectivity.
I believe the Provençal in this case, however, because of one din-
nertime scramble. During the summer in all of Provence there is
an unwritten rule that one eats every meal outside. Even with the
thunderstorms, the rain almost always passes in time for restaurants to dry the chairs and tables quickly right before dinner—a
serendipity aided by the fact that we always ate late (at least by
Colorado standards). One evening the thunderstorms continued unabated and intensely. Everyone ate at the same time at the
inn, as is common in many small French restaurants, so we all
waited for the storm to pass, but it only did so late in the night.
At the last possible second before dinner was to be served, all the
restaurant staff—including waiters, chef, sous chef, and bus person—rushed out, cleared the tables of their cutlery, and set up
dinner inside. A friendly buzz in the restaurant during the meal
belied the adventurous nature of indoor eating at this time of

Typical irrigation ditch coming off the North Fork. Much of the North Fork land is flood-irrigated. The rest of the water comes from wells.

Portable irrigation sprinkler used throughout the Vaucluse.

year. If this had been a usual occurrence, the good-natured thrill of having defeated the weather gods, shared by staff and diners alike, would probably not have occurred.

Climate is a collection and averaging of the often quite varied weather that occurs in a place, and these two places probably have weather as varied as one will find most anywhere. What this means, obviously, is that if you are a farmer or a viticulturalist, you cannot depend on the temperature to remain docile or the rain to fall. Little can be done about the temperature, but clever humans have learned to augment the rain quite efficiently. Both places practice extensive irrigation, although the North Fork and western Colorado rely much more heavily on irrigation than does Provence.

This is the case in part because in most wine appellations in France, strict wine production rules prohibit the irrigation of grapevines. This ensures a "natural" product and probably also ensures ulcers for the viticulturalist. In the Côtes du Luberon appellation, however, because of the vagaries of climate, grape growers can irrigate until Bastille Day (July 14). This date appears to have been chosen more for symbolism than as a result of any careful climate analysis. Nonetheless, irrigation is an acceptable practice here as opposed to many other appellations in France. Other crops are also irrigated, but they do not share the strict regulations regarding when, where, and how much water to apply.

In the North Fork nearly every crop needs to be irrigated, including grapes, and no comparable regulations prohibit vineyard irrigation. Some irrigation is from center-pivot irrigation sprinklers and some from movable pipes, both of which draw water from underground aquifers. Most farmers and ranchers in this area, however, use surface water running through canals and acequias that get their water from the Grand Mesa and other surrounding mountains. Skill is required to align the grapevines or other crops so the running water flows down the rows and wets each plant equitably. Irrigation in the North Fork Valley is literally the lifeblood of farming and ranching. Without it this place would look much more like a true desert than a quilt of vibrant fields and vineyards.

Despite human efforts to mitigate climate through artificial means such as irrigation, wild swings in weather are normal for both the Coulon and North Fork Valleys. I seldom expect to find poetic allusions in Chamber of Commerce propaganda, but the Hotchkiss chamber's website surprised me with an apt and lyrical phrase when it stated that in the North Fork Valley during March and April, "winter dances with spring." What an elegant way to say that the seasons ebb and flow from the gorgeous to the annoying or even brutal and back again. One morning dawns a beautiful spring day, and by nightfall there are six inches of snow on the ground. Fortunately, by the next day the snow is probably gone, and the crocuses have again emerged from under their white blanket.

Luckily for both regions, the gorgeous weather in any season far outweighs the brutal. Winter days are often spectacular; fall weather, with its rich colors, is the best; spring is welcomed with great expectation and enthusiasm; and summer's heat and sun ripen the grapes and make the melons sweet.

Each of us carries our own ideas, prejudices, and understanding of a place—particularly the place we call home. When we go somewhere new, we experience that new place through the tinted glasses of our past experience. Maybe my own glasses have been colored by my forty-plus-year love affair with Colorado, but I see and feel many similarities, albeit with some significant differences, between the landscapes of the Hotchkiss area in western Colorado and the Vaucluse/Luberon region in Provence. These similarities and differences are what make individual landscapes interesting, what give them meaning to the person experiencing them.

THE LAND

---

*That land is a community is the basic concept of
ecology, but that land is to be loved and respected is an
extension of ethics. That land yields a cultural harvest
is a fact long known, but latterly often forgotten.*

—ALDO LEOPOLD, *A SAND COUNTY ALMANAC*

WHAT STRUCK ME SO POWERFULLY that morning in Hotchkiss and sent my mind flying to Provence was the way the land looked and how that view affected my memories and geographic instincts. There was the long east-west valley below me, with its season-affected river; there was the sere natural landscape that became lush and prolific only if water was added; and who could miss the elongated mountains running parallel to the valley, which, in turn, paralleled the river in its flow westward? Then there was the undefined, visceral feel of the place that transcends words but is vividly palpable nonetheless.

At this point an astute reader will rush to an atlas of each of the two places and locate them in the larger context of continents, oceans, and latitudes. He or she will find that the latitudes of the North Fork and the Coulon are significantly different. The North Fork straddles the 38°, fifty-minute parallel of latitude, while the Coulon is much farther north and sits astride the 43°, fifty-three-minute parallel—putting the Coulon more than 350 miles north of its Colorado counterpart. This is about the distance from Chattanooga, Tennessee, to Columbus, Ohio, which many would consider two different worlds. But the North Fork sits in the middle of a huge continent, and the modifying effects of large water bodies such as oceans and seas are minimal, if not nonexistent. The North Fork is also surrounded by high mountains that greatly affect the moisture and temperature regimes. The Coulon, on the other hand, is on the much smaller European continent and is dramatically affected by large water bodies, specifically the Mediterranean Sea and the Atlantic Ocean and—more important in this case—by the North Atlantic Drift. This drift is an extension of the Gulf Stream arising in the warm Gulf of Mexico and flowing northeast through the North Atlantic. It literally brings warmth and much more moderate climates to the entire western side of Europe, including much of France.

In spite of the latitudinal disparities, the valley-mountain topography is very much the same in both places, yet it also has considerable differences—like two diamonds made from the same carbon but with unique facets. For example, the valley floor of the North Fork is 4,500 feet (1,370 meters) higher than the Coulon Valley bottom. This is a big difference in elevation, but, while causing some significant temperature differences, it is hardly noticeable—that is, with the exception of people coming directly to the North Fork from sea level, who will feel the elevation change. The regional relief, or the elevation between the highest and lowest points in the region, is almost exactly the same in both places, however—5,500 feet (1,676 meters) for the North Fork and about 5,900 feet (1,798 meters) for the Coulon. The high point near the North Fork is Mount Lamborn at 11,396 feet (3,474 meters), and the entire Grand Mesa on the northern flank of the valley sits at about 11,000 feet (3,353 meters). The high point near the Coulon is Mont Ventoux (6,263 feet [1,909 meters]) just to the north of the Vaucluse uplands on the northern side of the valley.

In case anyone thinks that Mont Ventoux, at 6,263 feet, is not much of a mountain, let me state unequivocally that this is one of the steepest and most challenging stages in the torturous mountain climbs of the Tour de France. I think the riders in the Tour would agree that this is indeed a substantial mountain when they near the top and their leg muscles burn unrelentingly. One dramatic aspect of Ventoux's character that affects bicyclists and other living things is the mistral wind that blows across the peak on its way south from the Alps. The highest wind speed ever recorded at the summit was 193 miles/hour (320 kilometers/hour). That is very respectable when compared with the world's longtime top recorded wind speed documented at Mount Washington, New Hampshire, where the wind was once clocked at 231 miles/hour (372 kilometers/hour). (The new world record recently recognized by the World Meteorological

*Long-range view from the west toward Mount Lamborn (*left*) (11,396 feet [3,474 meters]) and Landsend Peak (*right*) (10,806 feet [3,295 meters]).*

Organization was a wind of 253 miles/hour recorded during Cyclone Olivia on Barrow Island, Australia.)

What was less overtly visible that morning but quite obvious was the deep commitment most of the landowners, the locals, had for their place. No one could look at the magnitude of effort, resources, and thought that went into the enterprise of intensively working this land without recognizing the deepest dedication—the long, straight lines of old apple orchards,

Looking north at Mont Ventoux (6,263 feet [1,909 meters]) from the top of le Petit Luberon.

the peach trees that have lived for generations, and the relatively newly planted vines of noble grapes. Successful work on the land also comes from an understanding of the things that make the land an ally and those that could make it an antagonist. Leopold's "cultural harvest" of caring for the land derives, at least in part, from a knowledge of where that piece of ground sits in relation to everything else that affects it from the natural world. The locals I met knew their spot on the earth like a longtime lover. Their knowledge of the climate, the bedrock geology and geomorphology, the soil, the native flora and fauna oozes from their pores—maybe not in explicit facts, phrases, and scientific nomenclature but in how they work the soil. The careful placement of the vines, the effort taken in creating and caring for the irrigation structures, the meticulous husbandry of the crops are all vivid and common examples of the locals' expertise and perseverance. These outward signs of how people work the land are good indicators of how they feel about their place.

When I go somewhere for the first time, I automatically start attempting to figure out what makes the place tick in the natural sense. This is not a well-thought-out effort that I undertake intentionally, it just happens. Instead of taking pleasure in human company or the delights of food and drink, for example, I am often distracted by the land and by what is going on around me. This involuntary endeavor is not always accurate or reasonable. My first impressions are often wrong and need considerable revision as I learn more. But this inveterate looking and wondering has its rewards, even if it does come from unconscious effort. When I first came to the North Fork and the Coulon, I kept to my geographic instincts and started looking for/at what the natural milieux of the places were.

What I found were two places thousands of miles apart that, with obvious variations, could be twins—fraternal twins, but twins nonetheless. The foundations of both valleys are made up of rock fashioned in different ways but molded by time and geologic processes that are universal. These rocks are mostly sedimentary, and the unconsolidated mineral matter (sediments) of the valleys had their origins eons ago. In the case of the Coulon, the rock is almost ubiquitously some form of limestone—the calcium- and carbon-based rock developed in shallow seas over the course of millions of years. Nearly the entire area is underlain by some form of *calcaire* (for example, *calcaires lacustres,* or lake-developed limestone; *calcaires bioclastiques,* with small calcified plant and animal remains visible; *calcaires argileux,* with a clayey texture—you get the idea). These limestones weather into fertile soils that have physical and chemical differences depending on the clay or sand content, size of the gravels embedded within them, and water availability. In the valley bottom these limestone-derived soils are augmented or perhaps even totally covered by extensive deposits of finer-grained alluvium (stream-worked sediments) eroded, transported, and deposited from material brought down from the bordering highlands. Both of these major classes of soils are a dream for farmers and vintners as long as they can get the water to make them productive. The valley itself is covered by many feet of alluvium, but the limestone slopes away from the river are covered only by a thin veneer of soil. In fact, there is so much limestone that one is not far from a limestone quarry anywhere in the valley. Magnificent outcrops of this stone can be seen in places like the village of Lioux and its imposing Falaise (cliff) de la Madeleine. One of the best views of this wall of rock is along the road from Joucas to Murs, about a kilometer north of the village. This is a striking scene—in the

The spectacular Falaise de la Madeleine above the small village of Lioux. This is one of the best views of the massive limestone deposits within the Vaucluse.

midst of the bucolic orchards, vineyards, and fields, this monolith rises from the ground and looms over the small commune of Lioux. It would be eerie to live in the shadow of such a monstrous piece of stone.

The vast majority, although not all, of the foundation bedrock in the Coulon is limestone. There is a small yet significant area called the *ocreux d'Apt*. These are sandstones mixed with some clays that have a reddish to pinkish pigment that comes from the oxidizing (rusting) of iron-saturated minerals in the rock. Roussillon is the center of the almost defunct ochre industry, which processed and concentrated these pigments for industrial use. Most of this ochre went into staining the rubber used for items such as seals, washers, and tubing—the pink-tinted rubber products hanging on the walls of our local hardware store. Carole and I had been hearing about ochre ever since we set foot in the valley, so we visited the Conservatoire des Ocres just outside Roussillon to get the real story. As it happens, it takes a lot to extract and process the coloring from *les sables ocreux* (the ochre sands). The elaborate alchemy begins with the mining of the ochre sand. It then continues through washing, mixing, separating, settling, cutting, drying, cooking, grinding, sieving, and, finally, shipping. All this for a dark, burned-red color to make plumbing supplies attractive. I think I know why the ochre industry is no longer very viable. But the rock it came from is certainly beautiful and eye-catching.

One day Carole and I made a short excursion to the area near Rustrel, the location of a geologic park called le Colorado Provençal—the name certainly got our attention. It turned out that this was a miniature version of a famous park in Colorado

Limestone quarry seen looking southwest from the village of Ménerbes.

Springs, the Garden of the Gods. There were vibrantly hued red rock chimneys (called monuments in western Colorado), cliffs, and, in very graphic French, *mamelons* (nipples). The formations were mostly reds and pinks made up of ochre similar to that in Roussillon, but other pastel colors finished the palette. It was a striking reminder of the connections between our two places and made us think fondly of the Garden of the Gods and the monuments and canyons of the Colorado Plateau fifty miles west of the North Fork.

There are few, if any, red beds in the local area of the North Fork. Most of the bedrock along the Colorado fraternal twin valley is a formation called Mancos shale, a fine-grained, soft rock that often weathers into a sticky, nutrient-rich soil. As with the Coulon Valley, here much of the bottomland is covered in alluvium coming down from the uplands—in this case especially from the West Elk Mountains to the east, which are almost entirely volcanic and drained by the North Fork. But the North Fork geology is a bit more complex than its French counterpart, with various sedimentary rock types overlying the Mancos. They include the Mesa Verde formation, from which coal is mined in places up-valley from Paonia, such as Somerset and Bowie. Another, very extensive volcanic area on the northern flank of the valley caps the Grand Mesa. The top rock stratum of the Grand Mesa is an ancient basalt flow that hardened into a resistant rock not easily eroded, which caps the mesa and makes it more difficult to wear away. When basalt does weather, it becomes very fertile soil. Many of the smaller, lower mesas below the Grand Mesa—such as Rogers, Garvin, and Pitkin Mesas—benefit from the deposition of gravelly deposits coming off the Grand Mesa. Because of the gravel these soils are difficult to work, but with a little water and a lot of backbreaking effort

The coarse gravels on the mesas that flank Grand Mesa are perfect soil for growing quality grapes.

to clear the larger stones, they become incredibly fecund and superb for orchards, especially vineyards.

The perched villages of the Coulon are organic with rock—the stone of the buildings looks as though it exudes naturally and directly from the ground. On these hillsides above the valley floor, the rock defines the land. As one goes north and up from Joucas, for example, to the small commune of Murs, the limestone outcrops insert themselves into the living ecosystems in a landscape ubiquitously called *la garrigue*. Rock and tree, underbrush and fauna all become a dry and dusty landscape not particularly inviting to human visitors, especially in the midday heat of the summertime sun.

La garrigue is an upland of sparse, drought-resistant vegetation growing on thin soils and exposed rock outcrops. In this part of Provence, the principal ecosystems at the higher elevations are the *chêne blanc* (white oak) and *pin sylvestre* (Scotch pine)—neither of which grows impressively tall in this sere land. Scrubby vegetation such as the *chêne kermes* (kermes oak), with its spiny leaves, is interspersed among the larger trees. Other shrubs, grasses, and a few herbaceous plants fill in the landscape. At slightly lower elevations the white oaks are replaced by *chêne vert* (green oak) and the Scotch pines by *pin Alep* (Aleppo pine). For those not prone to recognize tree species, little will distinguish the two zones.

But the exposed limestone is really the story here. One day while hiking on a dry and gravelly trail on du Plateau des Claparèdes above Viens, about 30 kilometers (19 miles) east of Gordes, Carole and I encountered what we thought was an ancient rock hut. The dry stone construction and beehive shape of this *borie* reminded us of the eleventh-century *clocháin,* or beehive huts, we had seen in the Dingle Peninsula of Ireland. We later learned that an entire reconstructed Village des Bories was a tourist attraction near Gordes. We also learned that the dry stone structures here were built considerably later than their Irish kin, probably in the fourteenth century, and were used into the nineteenth century. The styles and building techniques were eerily similar despite the time and place differences. There are about 2,000 of *les bories* on the Vaucluse massif and many, as at Gordes, have been rebuilt from piles of stone and rubble. The uses in France are much less spiritual and ecclesiastical—they

A typical borie *of the Vaucluse. Most were built of unmortared stone in the last two or three centuries to be used by shepherds as shelters.*

were used primarily as storage sheds for tools and sheep. At times they were used as shelter for shepherds and as refuges for people caught out in the elements in the uplands.

More than anywhere else in our two valleys, the uplands are most alike—from afar they would be difficult to distinguish from each other. The *garrigue* in Provence becomes the piñon-juniper woodland of the hills and mesas of the North Fork. The exact species of trees may differ (or not in some cases), but the forest structure, the look, the very smell of the landscapes are acutely akin. The two main tree species in the Colorado landscape are the piñon pine and Utah juniper. Neither grows particularly tall or majestic, and each grows slowly in this dry land. They give the ecosystems their duplicate look and feel to match the white oak and Scotch pine forest of *la garrigue*. The only significant difference to me is the gastronomic advantage in Colorado of being able to gather and eat the piñon nuts—a staple of southwestern cuisine and a major food source for locals over the last few hundred years. Unfortunately, the piñons are under siege during this climate-stressed period. The *ips* beetle has killed or weakened millions of piñon trees in the US Southwest. So far only a few trees in the North Fork have been affected, but the drier-than-normal conditions over the last decade have made them vulnerable to the *ips* and other pathogens.

The two valleys are nearly as similar as the uplands. In both cases, little is left of the original natural, native vegetation. Although the Coulon Valley has been more intensively cultivated over the last 1,000 years, the majority of the North Fork has also been transformed into agricultural croplands, pastures, orchards, and vineyards. Only along the very narrow strip of land adjacent to the river bottoms, called the riparian zone, do we find a more natural set of plant communities. Fremont cottonwoods, white willows, green ash, and the obnoxious, non-native water thief salt cedar (or tamarisk) all grow vibrantly—some too vibrantly—along the North Fork and its tributaries. The exotic, invasive tamarisk is rapidly becoming a real hazard to western waterways. It uses huge amounts of water and toxifies the soil with high concentrations of salt. Over the last two or three decades, it has made a significant dent in the numbers of new cottonwoods and willows able to remain viable. The Coulon has its own riparian collage of vegetation—including the *peuplier blanc* (white poplar), *saule blanc* (the same species of white willow along the North Fork), and *saule cendré* (gray willow)—which grows along the river and its tributaries to create a dense, shrubby mix. In both cases the lushness along the streams provides rich habitat for fauna that depend on the thick growth for food, shelter, and migration corridors. Shade is another commodity enjoyed by humans and animals alike that is in great supply in these sun-shy, cool, streamside oases.

One of the two areas' major similarities is the pervasive lack of water, especially water available during the growing season. Provence sits at the heart of what is called a "Mediterranean" climate. This delightful phrase means that the summers are hot and dry, the winters are cool and wet (at least wetter, if not truly wet). This causes us to pause in thinking of this area as dominantly agrarian because most of the precipitation falls during the non-growing season. During our times in the Coulon Valley, a few afternoon thunderstorms cooled things down on a very hot summer *après midi*, but little useful moisture was added to the dry soil. Because of the high temperatures, any moisture

Riparian areas, those that run along rivers and streams, are clearly evident all along the bottom of the North Fork.

The Coulon Valley also has riparian zones, but they are mostly confined to the larger streams and uplands.

that did fall evaporated quickly. The official weather measurements recorded by Meteo France show that the average July precipitation in the Vaucluse is 3 centimeters (about 1.2 inches). The highest average monthly precipitation is 9.5 centimeters (about 3.8 inches) during October. The average for the entire year in this part of Provence is only about 61 centimeters (24 inches)—enough to grow most produce if it fell when the plants needed it, but it does not. With July's high temperatures at least in the 90° F range every day, 1.2 inches of precipitation is hardly enough to wet the surface of the ground, much less grow the verdant crops seen everywhere in the valley. The North Fork Valley has almost exactly the same average July precipitation, at 1.04 inches. Unlike Provence, this is not a Mediterranean climate, and the place has lower precipitation year-round. The maximum amount of precipitation is also in October, with about 1.6 inches on average, but the yearly average is a paltry 15.8 inches, which cannot begin to grow the kinds of crops seen there without additional water. In climatologist jargon these numbers make this a semiarid steppe climate—not a siren call to farmers.

The great author Wallace Stegner wrote a classic book about the farmers-versus-aridity dilemma in the US West called *Beyond the Hundredth Meridian*—that line in the sand runs through central Nebraska and is a surrogate for the twenty-inch isohyet. Most agronomists agree that it approaches folly to try to consistently grow crops, even drought-resistant ones such as wheat, in a place that gets fewer than twenty inches of precipitation per year, without the aid of supplemental irrigation. If one is near that number, one needs the moisture to come during the growing season, if possible. Obviously, neither of these valleys should be in the agronomy business. So how is it that each, in its own way, is a veritable garden?

The first time we went to the inn in Hotchkiss, we took a short hike down into the deep valley just to the north of the main house. We were accompanied by Leroux, one of the trusty dogs that "escorts" visitors on such a walk. We scrambled down the stony, dusty hill among the seemingly bone-dry piñons and junipers and soon heard what sounded like gurgling water. As we descended further the sound got louder until finally we were standing next to Leroux Creek, a perennial stream with its headwaters high above and to the north on Grand Mesa. We knew the Leroux Creek Inn was named after this stream, but as longtime Coloradans we expected some small trickle of water at best or, more likely, a dry streambed that ran only after a substantial thunderstorm or during the spring snowmelt season. Instead we stood by this impressive stream that was ten to twenty feet across and two feet deep—in most of Colorado it could easily be called a river.

Because of the general shortage of water, Colorado has developed some seemingly obtuse and restrictive water laws. The Colorado Doctrine, or law of prior appropriation, is the method the state devised in the nineteenth century to deal with this lack-of-water problem. In most places east of the 100th meridian in the United States, a riparian rights water law is in place. This general type of law makes some sense. It says that if you live on the banks of a stream or river, you can use the water in that stream for irrigation, drinking, washing your car, whatever. In Colorado, however, this is less straightforward. You are appropriated water for some "useful purpose" on a first-come-first-served basis, even if you do not live on the water source, as long as you can get the water to you. The rights to this water are regulated by the complex concept of the prior appropriation doctrine, which basically says that the first person who gets and uses water from a source has the highest priority to that water. The next person who does the same will have the second-most-senior rights, and on down the line. These water rights remain and are owned by you or your heirs forever as long as the water is used productively—however that is interpreted. These water rights can also be bought and sold like any other commodity. The oldest of these water rights in Colorado dates back to 1851 in the little town of San Luis, coincidentally the oldest town in Colorado. The rights are watched over and managed by a state government agency organized by major watersheds. Nearly every drop of surface water in Colorado belongs to someone, and in a water shortage period, those with the oldest rights get their full complement of water first. If in a given year of drought there is not enough water to satisfy all the rights, some places go dry while others get their full allotment. The situation becomes even more complex because Colorado is a source of water for other states downstream, and federal laws require the state to leave enough water in the major river systems to satisfy interstate water compacts.

The sound of mountain water we heard in Leroux Creek on our hike was the sound of someone's livelihood, or, more cynically, the sound of mountain money. The Leroux Creek vineyards are watered by shares of that creek; and every other farm, orchard, and vineyard owns a certain number of shares in one creek or another, or even in multiple creeks. One share can water approximately five acres of a vegetable crop for one growing season. Almost all irrigation in the North Fork Valley is accomplished with surface water versus the groundwater used in other places in the state. Both large and small irrigation canals and ditches run everywhere in the valley, and most of the backbreaking work of farming here involves managing water flows for the precious shares. Almost every map of the valley shows ditches and canals to move this water to the share owners. To help control the water, reservoirs have also proliferated. Fruit Growers Reservoir, Baxter Reservoir, Paonia Reservoir, Patterson Reservoir, and a multitude of nameless ponds store, regulate, and dole out the precious liquid.

A Coulon Valley map could almost be mistaken for one in western Colorado, with its mind-confusing array of canals, springs, fountains, streams, wells, cisterns, and water tanks. There is some groundwater irrigation, but, as in the North Fork, the vast majority comes from surface flows coming off the high lands that parallel the valley. Many of the streams in the Coulon appear from the rock like magic. Because there is so much limestone, there are uncountable caves and grottos where water flows underground, only to emerge as a full-fledged stream at the base of a cliff rather than as a small seep, as is the case with most springs in western Colorado. One of the region's famous little villages is Fontaine-de-Vaucluse where la Sorgue Rivière gushes from a large grotto at the bottom of a 300-meter (1,000-foot) cliff. The village has a spa-like atmosphere, and the cooling air next to the stream and the shade of the deep gorge are welcomed by the hordes of tourists walking along the lovely banks.

Colorado law is a bureaucratic puzzle, but France has some of its own water conundrums. As stated in chapter 1, most wine appellations in France do not allow supplemental irrigation for the vines in their purview. Perhaps they view it as cheating to regulate something as natural as precipitation—the grapes and the vintners must suffer to make good wine. But in the Côtes de Luberon appellation, a vintner can irrigate until July 14— Bastille Day. Few, if any other, AOCs (*appellation d'origine contrôlée*) in France allow irrigation. But without it, this very dry land in the Vaucluse would be unable to support a viable, quality wine industry.

The future of water availability for both of these places holds no good news. According to all climate models and an overwhelming number of scientists, the earth will warm substantially over the next few decades. This means that more water, not less, will be needed for almost all crops because of increased evaporation and plant respiration. In addition, precipitation patterns of distribution will be altered so that many places will get less moisture while others receive more. Most models predict that both valleys will have much warmer temperatures and the same or less precipitation. The stresses of the gambling lifestyles of farmers and vintners in both regions will undoubtedly increase. Perhaps the irrigation of vineyards will eventually become acceptable in other regions of France.

The lands, especially the higher grounds, of the North Fork and the Coulon would be considered marginal at best for any significant agrarian endeavor were it not for the human influences that have altered and molded the land. The soil is naturally fertile if water can be found in sufficient quantities and applied in a timely manner. But the elevated crowns above both valleys remain basically in their natural state. The rugged, stark *garrigue* and the piñon-juniper woodlands of scrub oaks, stunted pines, and fragrant junipers sit as a reminder to those living here that nature is not always benign and forgiving. The people have wrested agriculturally productive landscapes from a tough competitor, and that competition will not abate in the years and decades ahead.

VILLAGES

—◆—

Every day or two I strolled to the village to hear some
of the gossip which is incessantly going on there . . . taken
in homeopathic doses, [it] was really as refreshing in its
way as the rustle of leaves and the peeping of the frog.

—HENRY DAVID THOREAU, *WALDEN*

THE PHYSICAL GEOGRAPHY OF PLACE is the indisputable foundation upon which I have constructed the common vision of two like landscapes. But the human concentrations in the small towns and villages add a critical aspect to the two regions that is essential to our understanding and appreciation of their similarities and differences. The small towns of the North Fork are understandably dissimilar in many ways, although these differences are like a set of études on a single theme as opposed to individual works. The same can be said of the villages of the Coulon. Each French village is unique in detail, but the differences are only variations in degree. This discussion of the "urban" environments will cover nearly all of the towns loosely included in the North Fork region and a good representation of the villages in the Coulon (see the maps on pages 13–14 for the locations and distribution of these peopled places).

The first time Carole and I traveled to the North Fork Valley, we drove there along the beautiful ribbon of Highway 92, which loosely follows the meanders of the north rim of the Black Canyon of the Gunnison. This sinuous, seemingly unknown and unpopulated route is one of my favorite drives in all of Colorado. The road is scenic but not in the manner of many Colorado roads that run below massive, snow-blanketed peaks or above treeline in the alpine wind and sky. I love those high mountain journeys, too, but Route 92 has a special texture and feel that are more like an old-fashioned quilt made for a fall festival than a goose down comforter sewn for cold winter nights. Certainly you can see the high mountains of the San Juans to the south as they create the horizon, but Route 92—which runs around the West Elk Mountains' southwestern flank—is an unparalleled bazaar of vegetation communities, rock landscapes, and very few people. In the autumn the colors along the highway are highlighted not by the golds of aspens but by the more New England-y reds and oranges of Gambel oak, sumac, and mountain mahogany with enough aspen thrown in to remind us that we are still in Colorado.

After nearly an hour of driving generally west-northwest from Blue Mesa Reservoir, we were approaching Crawford and the surprisingly dramatic landscapes of this cattle town punctuated like an exclamation point by Needle Rock, a 35-million-year-old volcanic neck. Highway 92 meanders its way through Crawford and past the sturdy stonework fronting the newly recycled Community Center, then out of town over Crawford Mesa until it starts its descent into the North Fork Valley. As the route crosses the North Fork itself, it enters Hotchkiss and soon intersects with Highway 133 coming from Paonia to the east.

All the towns in the North Fork Valley have a common genetic code—agriculture based on good soils and sunshine, settled during the early statehood days of western Colorado. Any structures over 100 years old in this part of the country are considered ancient, and these towns qualify for senior citizenship on that account—they were all begun in the 1880s or so as service centers for the nascent farming and ranching that are such a big part of the land even today. This is not meant to discount or trivialize the thousands of years Native Americans used this place without much permanent visible impact. The villages in the Coulon Valley, on the other hand, are an order of magnitude older—they are well over 1,000 years old in many cases. For example, Gordes dates back to the fourth and fifth centuries (named Vordes by the Romans), and there is evidence that Ménerbes goes back to prehistoric times.

No one would confuse which towns are in Colorado and which in Provence if shown pictures from the North Fork Valley and the Luberon or Vaucluse. Most of the towns in the French selection are hilltop villages with a preponderance of ancient stone buildings and narrow, winding streets. The North Fork Valley towns are typically composed of clapboard houses with a few brick buildings for wealthier landowners. Most of the Provençal hill villages are overrun with tourists during the summer months; the American towns less so, although they are

Needle Rock is a volcanic neck that rises above the small town of Crawford. The rock is an icon of the area and the subject of many local artistic efforts.

trying to catch up very quickly. Organizations such as chambers of commerce are making a concerted effort to bring in as many tourist dollars as possible. Although the physical construction and appearance of the towns and villages in the two regions are from different worlds, the residents are surprisingly alike; ignore the English-French language difference, and they would be hard to tell apart. Each place has a large number of farmers and a contingent of typical civil employees, but there is also a large collection of more "modern" professions in all the towns and villages. A quote from the Hotchkiss Chamber of Commerce site could be used for any of the towns in Colorado or Provence: "It's hard to swing a stick in this valley without hitting an artist, two massage therapists, and a yoga instructor." One might add that with the swing of the stick or perhaps a lasso one would also hit a cowboy, a sheepdog, two coal miners, a chef (or at least a sous chef) from France, and a barrel race winner. Both regions are experiencing an in-migration of well-to-do residents from near and far—Aspen, Vail, Denver, Texas, and California in the case of the North Fork and Paris, and other European Union countries—especially the United Kingdom—in the case of the Coulon.

The Colorado Towns

Hotchkiss is unpretentiously becoming the prototype for the evolving culture of the North Fork Valley. From all outside appearances it is a typical western farming community—one long main street, clean sidewalks along the main drag, 100-year-old buildings with sporadic new paint jobs, and the usual collection of small stores, cafés, family-owned groceries, and real estate offices. But on closer inspection it contains all the elements of the emerging cosmopolitan mix of new and old, high and low culture, and the spirit of optimism that permeates the entire valley.

A single issue of the *North Fork Merchant Herald*, a monthly paper published in Hotchkiss mostly to promote commerce and civic events, emphasizes this point. In the March 21–April 17, 2006, issue, the *Herald's* regular section on "Valley Arts News" has announcements on the new Creamery Arts Center opening with galleries, an art reception, an artists' yard sale, and a full agenda of art classes for kids and adults. Four pages later is an article on the Western Slope Environmental Resource

Downtown Hotchkiss with Mount Lamborn in the distance.

Council, an organization that works toward a sustainable future for the valley. The article includes a sophisticated yet localized definition of sustainability—"Ensuring the social well-being of our communities while securing our natural endowment of ecological integrity, native biodiversity, and wildness of our public lands." Seven pages later the "Cowboy Pome" section in the *Herald* includes Lazarus Washburn's (a pseudonym for the paper's editor) "Gout Festival":

> The festival committee
> For the small mountain resort town
> Met to plan
> The summer schedule including
> Three wine festivals
> A mushroom picking and cooking festival
> And a "slow foods" festival
> Followed a few weeks later
> By the annual Gout Festival.

Some of Lazarus's poems are funny, some political, and others just plain irascible. But they all reflect the independent kind of libertarian philosophy that has been a hallmark of the West for more than 100 years. This particular poem gives a tongue-in-cheek (maybe) analysis of the "new" community and its trajectory into the future. A healthy give-and-take tension seems to exist between the "old-timers" and the new arrivals that might bode well for the town's future. But an underlying fear for the area's future was expressed in an offhand remark made to me by one of the newer residents: "the valley is becoming home to all the millionaires that are being pushed out of Aspen by the billionaires."

Hotchkiss exudes a sense of small-town Americana, just as thousands of other towns do. But this first impression is to some extent a facade, and the town's future is going in a direction many communities around the country would envy. This is a conservative community, welcoming to all who care to come and love the place but wary of outsiders who might view it as the next "happening" spot or who come to show the locals how things should be done. It has a rural lifestyle that accepts and sometimes enhances the cultural amenities of more urbane places. The town likes the past but reluctantly embraces the future like a child with some exotic new toy—not able to entirely fathom what it is all about but curious enough to want to play with it and see what happens. The current locals, however, want to be sure the place does not become just that exotic toy for the next arrivals.

Except for Joucas, discussed later in the chapter, Crawford is the smallest of the towns discussed in the two valleys, with barely over 300 people listed in the 2000 census. Unlike Joucas, Crawford is a bustling small community with little tourism that prides itself on being a typical American "cow town." Crawford has only a few scattered bed-and-breakfasts, no hotels, and certainly no spas. The little town has been touched only slightly by the new lifestyles seen in the remainder of the North Fork Valley, especially in Paonia, Cedaredge, and Hotchkiss. When one drives in on Route 92 from the northern edge of the Black Canyon, the main street—called Dogwood Avenue—winds its way up from the valley of the Smith Fork River (a small tributary of the North Fork) through old stone and clapboard buildings. The upper level of Dogwood passes by the stone walls of the Crawford Community Center and soon leaves the western

The old stone community center in the middle of Crawford.

edge of town. If you drive slowly you will get through the town in less than two minutes. There is a bit of updating here, with a couple of revived restaurants and a new bridge going in over the Smith Fork, but it is still a ranch town with the feel of the West in every pore.

It would be hard to find a better example of small-town America, with its Fourth of July fireworks display over the lake just south of town, the requisite Pioneer Days in June, "Render the Rock" art day in September where local artists spend the day creatively rendering Needle Rock, and the Village of Lights celebration in November. Springing up with some regularity are small glass-blowing studios and blacksmith shops that create decorative ironwork. But Crawford's basic character is the same now as it was twenty-five to fifty years ago—a cattleman's working town set in the morning shadows of Landsend Peak and the West Elk Wilderness. This working town is significantly less affluent than the rest of the valley, and there are few farmers at the higher elevation of Crawford Mesa—certainly not the boutique greenhouses with micro-greens for the Aspen market. The 3° F to 5° F cooler temperatures, later frost in the spring, and earlier frost in the fall preclude any delicate crops such as fruit trees or vineyards. The main crop on this 6,500-foot mesa is hay for cattle feed, with hay fields spreading out from the town in all directions.

Crawford is a place new residents can come to and learn to love, even if it takes a while to establish relationships with the locals. Eugenia Bone has written a funny, poignant book about her family's move to Crawford from New York City. *At Mesa's Edge* is the story of inexplicable changes in a diehard New Yorker and her slow evolution to a gun-toting, fly-fishing, fruit-canning ranch wife in an alien place that becomes home, or at

64

least a second home that she loves. Those contemplating moving to Crawford should read this account very carefully before committing to such an adventure—and an adventure it would be. Then they should move as fast as possible because Crawford is that still-unspoiled kind of place that is sure to be discovered sooner rather than later.

Crawford exhibits the individualism and independence many expect of small western towns. One tiny artifact that reflects this independent spirit is seen on the official state sign announcing the entrance to Crawford from the west along Route 92. This sign states that the elevation is "6,800" feet (all Colorado communities have an elevation listed on the official Colorado

Hay fields that surround Crawford indicate that it is much higher than the farmland down in the North Fork Valley, where fruits and vegetables are the agricultural mainstay.

Department of Transportation road signs that announce town limits). But someone, in neat, bold, black numbers, took it upon him- or herself to change the elevation to "6,549"—obviously at least one local wants accuracy and does not want to deal with the state bureaucracy to get the sign changed by highway authorities.

Bloom Tours, Cherry Days, the Mountain Harvest Festival—this is the face of Paonia that is rapidly overtaking the view the town had of itself for decades. For many years the town's persona was one of a community immersed in the dirty, dangerous, extractive industry of coal mining. Many Paonians still make their living in the coal mines northeast of town, most near the small community of Somerset. The coal seams are in the Mesa Verde formation, the same formation seen all over western Colorado. Although coal mining has been Colorado's unheralded mining enterprise, it is or has been very substantial and critical in places such as Crested Butte, Hayden, Craig, and Paonia. But Paonia is steadily trading its gritty reputation for a new, semi-sophisticated identity. Its economy has always had an agrarian side; now, however, the environs around Paonia are becoming the fruit/organic produce capital of the North Fork Valley and maybe even of western Colorado. Like the farmers around Hotchkiss, many growers here are moving toward organic and natural production.

In addition to the extensive orchard business is the exponential growth of very small (most under 10 acres) organic farms producing everything from garlic and grapes to beets and bell peppers to popcorn and pumpkins. One wet fall day we visited several small organic operations run mostly by recently arrived locals (*recent* being a relative term for anyone who has lived here for fewer than twenty years). One such farm was the Zephyros Farm and Garden, owned by a young couple with small children. They have several greenhouses in which they grow herbs, tomatoes, and cut flowers mostly for markets in the Roaring Fork Valley over McClure Pass to the east. They also raise goats for making goat cheese—much like the cheese found ubiquitously in Provence. This couple is well educated, and they are very dedicated organic growers who see this lifestyle as a sort of farming crusade to create a healthy earth and a nurturing place for families. Another of these farms was the Small Potatoes Farm just outside Paonia on Lamborn Mesa. We bought beautiful large braids of organic garlic fresh from the field that lasted us for months. We went back the next year and were devastated that the farmer had sold all her garlic by the beginning of October. This small plot of land also had the most exquisite patch of chili peppers I have ever seen. There seemed to be at least a dozen varieties of every imaginable, and some unimaginable, color growing intertwined like a wild chili thicket. The muddy field exhibited the reds, oranges, yellows, greens, pinks, and purples of these peppers in a collage of organic heat.

New-age agriculture is not the only change to Paonia's economic life. The town became the home of the *High Country News* decades ago when it moved from Lander, Wyoming. The front page of each edition of this regional western newspaper proudly proclaims that it is the newspaper "for people who care about the West." The writing is the best on western issues anywhere; it is serious journalism with serious journalists who do not take themselves too seriously. I should be ashamed to admit that my favorite section is "Heard around the West," in which

Betsy Marston collected odd and funny anecdotes about western foibles. The section is now written by another journalist who has retained Betsy's acute western sense of irony. My favorite spots should be the real news stories, which are very good and substantive. I appreciate them, too, but "Heard around the West" is just too funny and quirky not to love. One recent anecdote led with the statement, "If you're going to visit a brothel in rural Nevada, it's definitely smarter to leave your daughter home." The serious articles are about serious problems, and the paper's philosophical bent is somewhat progressive in a community that could be considered more than a little conservative. But the town and the paper seem to get along just fine; in fact, the paper is a source of some pride to the people of Paonia.

Another new contribution to the changing face of Paonia is the Chaco Shoe Company factory and headquarters on the southwest side of town. Chaco products have recently become the footwear of choice for many western and active outdoors people. These sandals are (mostly) very high quality, made locally (to some extent at least) using high-tech synthetic materials, and manufactured in a sustainable way. In fact, sustainability is a company commitment. It seems out of place to go to the factory and see in the parking lot pickup trucks with the requisite gun rack in the back window parked next to the filled bike rack. This modern shoemaker is thriving, a vibrant indicator that the economy in the valley is changing.

Paonia's landscape setting may be the most spectacular of any of the towns in the valley. It sits under Lamborn Mesa and in the shadow of Mount Lamborn. Paonia is considered the gateway from the west to the West Elk Mountains and the Raggeds Wilderness area to the northeast. Going east-northeast on Highway 133 from Paonia takes you past several active coal mines, Paonia Reservoir, and up McClure Pass—that passage-way to the self-proclaimed chic-ness of the Crystal and Roaring Fork River Valleys. The towns of the North Fork Valley are rapidly becoming a destination for people going the opposite direction who want a connection to the land, a place where place matters. As with other towns in the area, such as Hotchkiss, the fear is that the flow of "newcomers" will change the face of this valley. In an article written for the September 14, 2009, issue of the *High Country News* about a place on the mesa two or three miles west of Paonia, Michelle Nijhuis expresses her connection to the land in a way that could be a credo for many who live in the North Fork: "Our power came from the sun, our drinking water straight from the sky, our vegetables from down the road. For me, the land provided the tranquility—and the low over-head—I needed to start and sustain a writing career. Though I traveled often, and enjoyed the time away, I always returned to the mesa with a sense of relief."

Cedaredge sits nestled in the narrow valley leading northward from the main North Fork Valley. The town is the entryway to Grand Mesa looming to the north. It sits at about 6,200 feet and is just at the upper limit of the most intensive fruit-growing part of the area. In fact just south of Cedaredge is Orchard City (hardly a city but rather a mix of expansive orchards mixed in with some development more akin to an exurban ranchette community that are part of the Cedaredge environs). One of the main water features in Orchard City is the Fruit Growers Reservoir, a seasonally temporary home to thousands of Sandhill cranes as they migrate through the area each spring and fall.

Cedaredge might be the most "touristy" of the North Fork towns. It has more restaurants and tourist shops than the other towns, as well as the only golf course in the valley. It is the largest town in the valley, with about 2,000 people, and takes seriously its role as the gateway to the Grand Mesa Scenic Byway. In autumn it has the largest open-air festival in the region, the Cedaredge AppleFest and Gala. Everything apple can be had at this festival, which takes over the entire downtown. Civic pride and entrepreneurship are manifest in the many city events planned throughout the year—in April there is Bloom Town, in May Heritage Day and the opening day for Pioneer Town, in July the Little Britches Rodeo, in September Color Sunday (which celebrates the turning of the leaves), the AppleFest in October, a Village of Lights celebration in November, and finally the Christmas Festival in December. These events attract locals, but besides the fruit industry, tourism is the community's most important economic engine, so all these festivals, days, and rodeos are meant to attract tourists from far and wide. No worries about being crushed by hordes of shoppers, though; this is still a small town, and the crowds are certainly noticeable but not overwhelming. All the events have the refreshing small-town, local feel of civic pride in their place.

Les Vaucluse Villages

The village of Gordes in Provence might be the antipode to Hotchkiss—after spending time in Gordes, I'm not sure the village is a real place. It feels more like the Disney version of a beautiful, unblemished French hill town. It is one of the most beautiful and picturesque villages I have ever seen. In fact, Gordes is the quintessential *"l'un des plus beaux villages de France"* (one of the most beautiful villages of France—an official French government designation). It reminds me of a gorgeous and very expensive Ming Dynasty vase that is very nice to look at but in which most people would be petrified to actually store something. The first time Carole and I saw Gordes, we entered the village from the northwest, and all we saw were stone-walled enclaves seemingly designed to keep people out and for others to hide behind. There was scant life in the town, unlike most French villages, which are alive and buzzing—albeit at a slow pace when the tourists go home for the winter.

The second time we went to Gordes we found out too late that it was market day. We had to park a kilometer away and pay a very steep fee. We walked through 100° F heat for fifteen minutes just to get to the village. Market days are always busy in France, but we were unprepared for the stone wall–to–stone wall mass of humanity inching toward the village with us. As we should have expected, the Gordes market was very upscale and smaller than markets we attended in other Provençal villages. One feature of the market that nearly struck us dumb, however, was the Native American dance group decked out in full headdress regalia and accompanied by the heavy drumbeat reminiscent of powwows in the American West. What a contradiction to the haut trendiness of the rest of the market.

Gordes has at least two saving graces in my mind, however. First, we had some meals in a local restaurant; the food was superb, and the service was as friendly and efficient as we could have wanted. We ate dinner outside on the patio, and the long-lingering evening sun, the cooling breezes on the hilltop location, and the buzzing of the bees in the flowers around the patio all added to the pleasure. The second saving grace is the view

Gordes is among the most famous of "l'un des plus beaux villages de France" *(one of the most beautiful villages of France) in the Vaucluse. It has a spectacular hillside setting and attracts multitudes of tourists.*

from the town over the valley below. Spectacular yet bucolic landscapes spread out beyond its hilltop for tens of miles from the southeast sweeping to the southwest. These intricate farm fields, orchards, and, yes, vineyards might be the signature scene for the Coulon Valley. It is very much an exquisite tapestry— even well beyond a tapestry. The scene is more a kaleidoscope of vibrantly colored glass that changes hue and pattern with the turn of the seasons. Nothing resembles or is the size of the mega-factory farms found in places in the United States. The fields, vineyards, and orchards range from a few to perhaps twenty-five or thirty acres in size.

Having said all that, there is a history to this village that belies the current stylishness. In her collection of memoirs about villages in this part of France entitled *Luberon—Traces de Mémoire*, Évelyne Jouval interviewed dozens of people who grew up in these hamlets many decades ago—before the Luberon was "discovered." For her piece on Gordes, she talked to people who remembered the place as a normal village made regionally famous by the skill and concentration of cobblers who worked there. People would come from many miles around to buy shoes and boots in Gordes—I can just imagine the scent of leather permeating the air. Ironically, the old leather shoe industry in Gordes is now extinct, while the modern version in Paonia is thriving. Gordes was also known for the stone that was quarried there, which was used to build the ubiquitous outdoor bread ovens still perched in a few backyards. Neither of these historical highlights of Gordes's past remains, and I can only wonder what the old interviewees might think of the Native American dances at the chic market of today.

The tiny commune of Joucas (population barely 300) has little of the feel of Gordes—a few stone walls, few people in the streets, very few commercial establishments. In fact except for the restaurants that belong directly to the commune's hotels and inns, there are no restaurants and only one small grocery shop the size of a large American bedroom. Although the town's economy depends directly on the tourist trade, especially travelers who stay a week or two mostly in the walled hotels, the village itself seems to belong totally to the locals. On a scorching day, only a few hardy souls use the small park, where shade from the plane trees makes sitting pleasant, or walk the narrow stone alleyways through the village. An enjoyable hour or so can be spent trekking through the steep pedestrian passageways in search of the many artist ateliers in the commune.

The *mas*, or old country farmhouse turned into an inn, in which we stayed had everything one could ask for—quiet, great food, hospitality, bucolic beauty, a swimming pool to make up for no air conditioning in the 100-plus-degree heat, and wonderful owners/managers. Le Mas du Loriot was about a kilometer north of the center of Joucas in the beginnings of the *garrigue,* or upland scrubby woodland of Aleppo pine and green oak. Small patches of lavender and other flowers attracted hundreds of butterflies and many more bees. The constant buzz lulled us into a mid-afternoon stupor in the heat. It was a laid-back yet intense French immersion that made us consider moving there.

One small yet vivid incident put our Loriot experience into clear perspective. At one of the dinners served on the patio, there were six or seven tables, all with couples. It was a small scene of total relaxation and contentment. As lightning flashed in the distance without us hearing thunder, a very large, black spider started her descent from the overhead trellis, moving slowly

The country inn called Le Mas du Loriot where we stay near Joucas.

down the strand of silk she was excreting. This ominous-looking creature would have had a group of American diners scurrying for the RAID® or a shovel or some other implement to smash the beast. Here, however, nearly everyone just sat and watched with calm interest as the spider hung around for twenty minutes or so. Everyone continued to eat the delicious food, drink the good wine, and talk between tables about the *gros araignée* and how beautiful and big it was. It seemed that nothing could disturb the tranquility and the joie de vivre.

In Jouval's collection of memoirs, the remembrances of Joucas conflict little with today's reality. The people who grew up in the hamlet recall that if they wanted anything they had to get it outside the hamlet—little to nothing commercial was available there. The place was surrounded by wheat fields, and the summer air was filled with dust from the combines harvesting the grain. Today, a few wheat fields and many vineyards surround the village, and grapes overshadow other crops in terms of area cultivated. Little else has changed in Joucas save the big, walled hotels, and maybe that lack of change is a good thing.

The Coulon Valley is somewhat bigger than the North Fork Valley and has a much wider and flatter valley bottom. There is also a significantly larger overall population there, with most people living in the small villages and communes, hamlets really. The main towns such as Gordes are all hill towns built of stone with ancient narrow alleys and passageways. Ménerbes is one of these vertical places that dominate the rim of the valley. It is an exceptional place even compared with the other "most beautiful villages of France," a true fortress town in the region, built on a high, protected promontory. It sits 90 meters (about 300 feet) above the valley on a spur of rock on the northern flank of le Petit Luberon Mountain. The thin rock fin juts to the northwest, and it seems it would be nearly impregnable to anyone trying to scale the heights while it was being defended. That was likely the thought of the Huguenots, who dominated the town during the Reformation in the sixteenth century. Provence is a particularly Catholic part of France, but it is also full of very strong-willed, independent people who defy authority in all its guises. The Ménerbes area was a regionally major zone of confrontation between Catholics and Huguenots in Provence; the conflict ended only after a prolonged siege of the Huguenot-held village.

Today, except for the exceptionally solid walls of the Citadel and the church, which remind visitors of the old fortifications, the village is relatively quiet—only blowing car horns and tour-

ists' loud voices disturb the peace now. Ménerbes is typical of all
these hilltop towns, which are caught in the jaws of a cultural
vise—keep the village small, quiet, and traditional or expose it
to all things touristy. The first time we went to Ménerbes, two
signs juxtaposed on the stone walls abutting the parking lot were
evidence of this vise. The first said (in English), "Don't leave
anything in your car." Crime is not much of an issue in this valley, but theft from presumably rich tourists is the exception. The
second sign (in French) said, *"La vie est belle"*—life is beautiful.
This sign was obviously hand-drawn and heartfelt, a touching
welcome to the thief-"infested" town.

There is a long stone wall on the southwestern edge of the
rock spur on which Ménerbes is built. It is a beautiful walk along
the wall that overlooks a gorgeous, intricate landscape in the narrow valley below. The view from Gordes is expansive and kalei-
doscopic; here the landscape seems to have been reduced like
a good French sauce to only the very essence of the place. The
fields are smaller than those in other parts of the region and are
apparently tended with infinite care. Even the limestone quarry
across the valley is picturesque, with its strikingly white stone
"steps" in high contrast to the mountain vegetation on the north
flank of le Petit Luberon.

One of the little oddities of Ménerbes is the corkscrew
museum (Musée du Tire-Bouchon) just one kilometer northwest of the village. Yes, the French do take everything about their
wine very seriously. Believe it or not, they have over 1,000 different types of corkscrews! Ménerbes also has the Maison de la
Truffe et du Vin du Luberon—the House of Truffles and Wine
of the Luberon. I am not a big fungus eater, but I was very much
drawn to the House of Wine idea. The beautifully restored old
hotel from the seventeenth century now houses a collection of

exhibits and a wonderful wine cellar devoted to the small Côtes du Luberon appellation wines. The woman who helped with our "research" was quite knowledgeable and helpful. We had a great time talking to her and choosing the right wines until she had to go rescue the wine racks from a loud American couple who were pulling bottles from the rack without help. They nearly dropped an entire case of very good wine onto the hard stone floor. A word of advice for those who have not shopped in a French village grocery, bakery, or wine shop: do not help yourself. This is a sure way to get yelled at in untranslatable yet very understandable French—I know this from personal experience. So for this woman the rescue job was an affront to her culture as well as a threat to her wine. We lingered after the other couple left to try to make amends for Americans and left with some wonderful bottles of the local vintage.

The southern, defining escarpment to the Coulon Valley is delineated precisely by the long mountain front of le Petit Luberon and le Grand Luberon Mountains—an east-west mountain front that runs for about 40 kilometers (about 25 miles) with only one possible pass, the Combe de Lourmarin, which is guarded by the village of Bonnieux. Long considered the first of the perched villages in the region, Bonnieux looks as though it was smeared onto the hillside when viewed from Lacoste about one and two-thirds miles to the west. Like all the other perched or hill villages, Bonnieux is a beautiful village with winding streets and stone-paved alleys that climb from the entrance of the village up to the *vieille église*, the old church, near the summit of the town.

The village is a tourist destination like all the hill villages of the Coulon. One morning while Carole and I were having a pastry and coffee at an outdoor café, only one other table was occupied. The other couple started a conversation with us. They were British with a capital "B" and probably talked to us because we were from a former colony—the man's father had been the police chief of Shanghai in some distant past, so the concept of colonies occurred to me. They started complaining about the slow service at the café in particular and the general French indifference to their British sensibilities. We were having a delightful morning sitting on the sun-speckled patio waiting for our order and were in no rush to leave, so their complaints were not reciprocated. We found the slower pace of service in

The village of Bonnieux from the uplands of le Petit Luberon.

restaurants and cafés calming and the kind of atmosphere we expected on a foray in a French countryside village. We eventually left this unhappy couple and began our leisurely drive over the Combe de Lourmarin into the broad valley of the Durance to the south. This kind of cultural insensitivity can also be found in western Colorado. The North Fork and other tourist destinations, in particular the ski areas, are often locales where people from other states and countries come who sometimes want a calmer pace and a more laid-back attitude. It does not happen often in France or Colorado, so when it does, such behavior certainly stands out.

Bonnieux had been a bastion of Catholic fervor during the Reformation and was the religious rival of nearby Lacoste, the equivalent of a Protestant fiefdom during that period. Lacoste today may be the most globally famous village in the Coulon Valley and is certainly more famous than any town in the North Fork Valley. Le Château de Lacoste is the ancestral home of Louis Alphonse Donatien comte de Sade, also known as the Marquis de Sade—yes, that Marquis de Sade. The chateau is now mostly a ruin, but the ruins have been stabilized and many tourists hike the steep streets up to the château or attend the yearly opera fest within the old walls.

Lacoste and the château have become the home of L'Espace Cardin, Pierre Cardin's adopted business and artistic home. The town has become famous again because of the Lacoste brand name Cardin has given to a line of his clothing. Cardin and the village have molded a creative center in Lacoste, with opera, theater, and other music festivals throughout the summer months. We could see the lights from these performances as we sat and ate on the spidery patio that evening at Joucas across the valley. In fact Lacoste has become an artistic epicenter. For example, the annual summer program of classes run by the Savannah College of Art and Design in conjunction with the Lacoste School of the Arts is a real draw to this small village. During these summer episodes, the entire village often becomes a piece of artwork. During one of our visits the theme was "*Les Chimères,*" or the fantasies. The filaments of this work were spun throughout the village like the work of another, even larger spider. It was an ethereal experience to walk the village's steep stone alleyways and see web after web in every corner and the art students hanging out doing what art students have done for generations—arguing, smoking, drinking, and trying to look chic.

Lacoste was not always avant-garde or steeped in sadistic eroticism; for most of its existence it has been a normal, even poor farm village with a normal farm and village life. Like Gordes, it had an active cobbler industry and typical spinner and bakery enterprises. One of the major events of the year was the annual Lenten play written and produced in the local patois—Provençal. This lyrical language has subsequently been overwhelmed by French, but it is still regarded with much local pride in the heritage—examples of which include the numerous road signs announcing the upcoming village written in both French and Provençal. According to Jouval's account, the biggest village affair in Lacoste over years past was a week-long village fête that culminated in an all-night ball. Longtime residents still relish the lavish costumes and funny antics associated with the fête.

Bonnieux guards the northern side of the Combe de Lourmarin; the village of Lourmarin sits at the base of the pass over the Luberon on the south side of the mountain. Although not in the valley of the Coulon and not a hill village, Lourmarin has been a vital link between the Coulon and the larger, even more agriculturally productive valley of the Durance Rivière. The Durance is a major river coming out of the Alps far to the east and flows in its convoluted, braided channel just a few miles south of Lourmarin.

Lourmarin is a beautiful village complete with a sixteenth-century château built on the ruins of a twelfth-century battlement. In fact, Lourmarin is numbered with others in the area as one of *les plus beaux villages de France*. The house we lived in while in the village was old, quaint, and beautiful. We had the lower level and access to the shaded terrace, which is normally a necessity during a Provençal summer visit. Our time, however, was mostly spent in the rain of a very strange, early summer weather pattern. Typically, summer is hot and dry. That year, however, the entire region had at least twice the normal rain for that time of year, and the city of Aix-en-Provence just a handful of miles to the south had three times its normal rainfall. When we walked to the center of the village, even early in the morning as I went to get our breakfast baguette, we were nearly always met by people at the cafés sitting outside in coats and under umbrellas smoking and drinking their morning espresso. It was wet, but the French (and most of the tourists) just went with the flow and ate outside, no matter how wet it was.

The village's three cafés are the center of activity. They start serving early (around 7:00 a.m.) and serve late (until after 10:00 p.m.), even though no food is served after lunch or *le déjeuner*—only wine, coffee, aperitifs, and the like are ordered. One late afternoon the young postman came tooling into the square on his yellow scooter with the La Poste insignia painted in blue on the side. He went to the *tabac* to deliver the mail and probably get some cigarettes and then came over to the café where we were having our own aperitif. He was still on duty by the look of

The entrance to our "house" in Lourmarin. It looks small but was spacious and quaint. Parking, however, was a nightmare.

all the mail on his scooter, but he greeted all the locals with the requisite double *bise* on the cheeks, sat down, lit a cigarette, and ordered a drink that came in a J&B scotch glass. French cafés are very particular that the glasses match the drink inside, so the J&B glass was probably not filled with lemonade. When he had finished his greetings, cigarette, and drink, he sauntered over to his scooter to finish his appointed rounds.

Part of what we love about France and Provence in particular is this casual joie de vivre. It does not depend on convenience, that god of American culture. In fact at times it seems there is an intentional, distinct, anti-convenience air about the area. A few examples will illustrate. The two pizza places in Lourmarin are both closed on Mondays, and when they are open you must order between 7:00 and 9:00 p.m. You can order a day ahead, but that precludes spontaneous plans for a pizza dinner. The village water in Lourmarin is fine to drink if you do not mind a mouthful of calcium carbonate from the limestone aquifer from which it comes. So bottled water is nice for a hike—assuming that the one *épicerie* (grocer) that is open has water that day (the other *épicerie* was never open during the time we spent in Lourmarin). There are no Internet cafés. To keep in touch with our family back in the United States we had to join la Bibliothèque (the library) for eight euros for the year. The library staff was helpful and friendly, but the library was only open fifteen hours a week—with some exceptions. We had to do a lot of precise scheduling to get pizza and e-mails. At times we got a bit frustrated but then took a deep breath and realized this was what life here is about—the adventure of everyday living and the go-with-the-flow attitude.

Perhaps the most intriguing village of all in the Coulon is the small ochre capital of France, Roussillon. Roussillon is intriguing because of what is known about its history, social life, and people over the last half century. In the mid-1950s an American sociologist from Cambridge, Massachusetts, went there to live and write a sociological account of life in the village. Laurence Wylie and his young family spent a year totally immersed in village life while he did intense observation of and research about what they were experiencing. Wylie's book, *Village in the Vaucluse*, was a very honest, often uncomplimentary text. To protect the villagers' privacy, he changed the village name in the book to Peyrane. Only after most of the original subjects had moved on through death or migration did he admit that the village he had studied and written about was actually Roussillon—although from descriptions and locations, many people had already guessed its true identity.

Wylie's account, first published in 1957, describes a village in some turmoil. Economic hardship had been and remained a constant. From the *phylloxera* outbreak that demolished the grape industry in the late 1800s, to the change to olive oil that was soon overwhelmed by cheaper oil from North Africa, to the rise and precipitous fall of ochre markets, to the economic and sociological rifts caused by World War II, to the near-loss of the school for lack of funds, Wylie's book discusses critically but with empathy a village unsure of its long-term survival.

The older locals grew up in a village of struggle and near-poverty. Water was always scarce in the summer, food was never plentiful, and the people had what could be described as a hard-scrabble existence—gathering greens from the roadside for dinner, using blood from chickens and rabbits to make omelets, bartering for wheat because no one had much cash. What a change a few decades can make.

To those visiting Roussillon today who have not read Wylie's account, none of these innumerable travails are evident. What is seen today is a spectacularly beautiful village built into the reds, oranges, and yellows of the ochre-tinted landscape. It has become a truly sophisticated Provençal village complete with a Michelin-blessed two-star hotel and multiple-fork restaurants. At least on the surface, it is one of the most economically vibrant and viable of the towns and villages in the Coulon. If I came into a small fortune and were able to move to this valley, Roussillon is probably the village I would choose. The old residents in Wylie's account would certainly not recognize the village today.

It is obvious from the narrative about the villages in Colorado and France that the towns in the two regions are different from each other. For one thing, the population density of the land in the Coulon Valley is about 125 people per square mile, while that in the North Fork Valley is approximately 15 people per square mile. The overall size of the respective counties/departments is about the same. The Vaucluse, the department in which all the villages in the French area are located, covers an area of 1,377 square miles; Delta County encompasses 1,149 square miles. The French valley seems bigger because speeds (at least speed limits) on the small French roads are lower and the roads are often much more winding and narrow than their Colorado counterparts.

The North Fork has fewer towns and villages by far. Each town and village in both valleys is distinct and individual, with its own stories and short or long histories. The towns I described for the North Fork Valley are the only real towns in the upper end of the valley, with just a few small, unincorporated settlements sprinkled in. They are all farming and ranching towns, with some of the new economies I mentioned developing rapidly. The perched villages of the Coulon Valley are dependent almost exclusively on tourism, and all of the more prosaic farm service villages are in the valley. These workman-like places include the villages of Lumières, Coustellet, Beaumettes, and literally dozens of other nondescript farming communities that tourists barely notice. The towns in the North Fork Valley and the farm communities in the Vaucluse are very comparable, each with farm implement stores, barns along the roads, and few tourists in sight—although more and more tourists are coming to the farm communities of the North Fork.

Finally, the villages of Gordes, Ménerbes, Lacoste, Joucas, Lourmarin, Roussillon, and Bonnieux are really, really old by any standard. Hotchkiss, Paonia, Cedaredge, and Crawford are kind of old by Colorado standards, but preservationists in Europe would not give them a second thought.

So why do I lump them into the same stew or ragout, if you like? They are both places that are attracting people priced out of other desirable places. Neither valley has the usual mountain attractions such as ski areas, so people must be moving there at least in part for the unhurried life of the rural landscapes. There is a spirit of entrepreneurship in both valleys that will help keep the economies robust into the future. But when it comes down to the simple answer, it is the people and their attitude of stewardship toward the land and their milieu that make the two valleys kin. The towns and villages of both regions are inextricably and directly linked to the land uses that lie outside their municipal limits. The people and their views of their milieu are the real essence of the places. If you talk to a native or a longtime resident in either country, the conversation soon comes around to food, drink, the beauty of the land, the vagaries of the weather, and his or her individual spot in the landscape—whether in town or on a farm. The people may not even realize that they exist as a living part of the place, but they surely do. They have a way of looking at themselves that has a connection to the place. This connection to place is what most of us would have needed just to survive fifty to a hundred years ago. But here in both cultures, the deep connections to the land and the towns are touchstones. Many of us have let that human character atrophy with our busy lives in a modern world or by being swamped by the daily deluge of public relations in the media. These places in Colorado and France reveal a life that slows down, looks around, and makes people feel connected to their place.

WINE

————————◦◦————————

I offer thee Mirèio: it is my heart and spirit,
The blossom of my years.
A cluster of Crau grapes, with all the green leaves near it,
To thee a peasant bears.

—FRÉDÉRIC MISTRAL, *MIRÈIO: A PROVENÇAL POEM*

Make me poor, I will make you rich.

—THE VINEYARD TALKING IN AN OLD PROVENÇAL PROVERB

SOME PEOPLE ACTUALLY READ those large, detailed tomes written about wine. These several-hundred-page volumes usually try to cover all major wines from around the world and look at every significant wine-producing region. Thousands of places need to be discussed, some in great detail and others only in passing, as wine is one of the most widespread and complex commodities produced globally. I admit, I am a wine nerd who reads these books—many of which are very well written, some even funny, and usually intensely informative. But never have I seen more than a quarter of a page on the wines from Coulon and never even a mention of the wines from the North Fork Valley. There are no Château Mouton Rothschilds or Château de Beaucastel Châteauneuf-du-Papes in the two appellations in the Coulon Valley separated by the river—the Côtes du Ventoux and the Côtes du Luberon. And absolutely no one would confuse the North Fork with Napa Valley—not a faux château to be found. Fame and size tend to drown out mention of smaller or newer or less glitzy wine regions. As far as I am concerned, that is a good thing. If the two landscapes discussed here were already famous for their wines, it would leave very little for this curious, thirsty geographer to discover. I am most assuredly not a wine expert, but I love wine and I love trying to understand landscapes, which is where these two disparate passions merge.

One might ask in all seriousness, then, what makes these two valleys and their wines so compelling? Put succinctly, they are compelling because they represent all the wines of the world that are not well-known by wine experts but that are really good and deserve to be better recognized. They are probably two among many hidden treasures the world's wine experts have ignored. Because they are virtually unknown, they allow wine lovers who are not experts and do not get free trips to tasting rooms, or *les caveaux*, to try the latest fad or newest vintage of some famous wine to explore the tastes and *terroir* of undiscovered territory. It gives me, and I hope others, the real feeling of adventure and secret delight in finding those small pearls that are generally ignored by others. So what have I found in these two enologically hidden valleys?

I will start with the Coulon because, unlike the North Fork and even though it is not much of a blip on the global wine radar screen, the tradition of wine in the region dates back thousands of years to at least the Romans, if not earlier. Vineyards have long been part of the landscape there and have made a permanent mark on the ground. *Les Vignes* are places on the Earth where nature and humankind have gotten together to create something special. Wendell Berry, that Kentucky-born philosopher of the soil, talks about the need to appreciate and preserve "well-used landscapes." By this he means landscapes that are used well. Vineyards are inherently examples of places that are well tended, lovingly cared for, and usually quite productive while preserving the essence of the place and the human culture attached to that place.

The Institut National des Appellations d'Origine, established in 1936–1937, is the organization that promotes and regulates the wine regions of France as well as other products such as cheese and olive oil. In the 1930s, vintners from around the country had for years been under a loose, confusing, and ineffective system of classifying and judging wine and the regions in which it was produced. Wine producers and promoters would do a lot of obfuscating when talking about *les crus* or *les grand crus* and noble grapes and on and on. Châteauneuf-du-Pape in the southern Rhône was the first wine area to establish rigid quality control; it sanctioned only certain grape varieties for a wine that used the name. From this beginning, a hierarchical system of controls and standards was created. Wine quality does not always follow this hierarchy as some might like to think, but it gives the buyer minimum assurance of what is being produced and purchased. Generally speaking, the ascending order of this classification of wine quality in France goes from *vins de table* (table wine) to *vins de pays* (wine of the region) to *vins délimité*

de qualité supérieure, or VDQS (high-quality wine of the region), to the top *appellation d'origine contrôlée,* or AOC. The AOC designation has the most control and the highest standards, and most VDQS areas are currently being upgraded to the AOC designation—the VDQS will be phased out over the next few years. At least that is the French plan, but then so was the Maginot Line. Believe it or not, this is the simplified code for French wines. This relatively new system is much better and more understandable than what was in place formerly, which was basically regional or even local classification chaos.

This discussion of the French appellation system illustrates where the wines of the Coulon fit into the French scheme of things. The Coulon Rivière cuts the area into two separate wine regions, or AOCs. The Côtes du Ventoux is north of the river and runs all the way to Mont Ventoux and somewhat beyond. It did not become an AOC until 1973. The Côtes du Luberon is across the river to the south and stretches all the way to the Durance. It did not get its AOC status until 1988. To the French wine mind, these two AOCs are little more than afterthoughts to the "great" wine houses and regions. My best guess is that many French wine connoisseurs and quite a few wine drinkers around the world have never heard of either AOC. I reiterate that this is fine with me (although *les vignerons* of the region might disagree).

The term *vigneron* is literally translated as "wine grower." Of course, people do not grow wine, they grow grapes. But when one thinks about what goes into a wine, the term really is quite accurate, if not literal. The wine is in the grapes—no good wine will be made from mediocre or bad grapes. And the quality of the grapes depends on so much. Most decent to great wines come from the grape species *Vitis vinifera,* but there are literally

The view from the second-floor window at the Leroux Creek Inn. This expansive view was the inspiration for this book.

The ochre cliffs of Roussillon were exploited for decades to produce red coloring, mostly for rubber products.

Striking red sandstone and shale formations at le Colorado Provençal are very similar to, if smaller in stature than, many red sandstone formations in Colorado.

A colorful, chaotic pepper garden that is part of a larger farm in the North Fork Valley.

This view of the landscape immediately south of Gordes illustrates the quilt-like character of the intimate landscape in the Coulon Valley.

A small vineyard seen from the hilltop village of Ménerbes.

Lourmarin is another of the beautiful villages of the Luberon/Vaucluse. Albert Camus is buried in a modest grave in the village.

The deep purple chambourcin grapes at Leroux Creek are almost ready for harvest.

Typical café collection in the middle of a village where several cafés serve in the square. At times the only way to know which café will serve you is by the color of the chairs at the tables.

Flax-like flowers cover large areas of the garrigue.

Requisite photo of lavender fields marching toward a hilltop village. In this case the village is Simiane-la-Rotonde at the far eastern end of the Coulon Valley.

These red poppy fields are spectacular and widespread. Many can be seen from miles away from the tops of some of the limestone cliffs.

hundreds of varieties of the *vinifera* grape to choose from. The start of a great wine is the selection of the variety or varieties of grape that will be used and that will grow best in the local region or on some micro-plot of land. This is no easy decision, since it may take a great deal of time, effort, money, and several years for the vines to produce to the point where the grapes are usable. Certain grapes such as the pinot noir and chardonnay tend toward the cool, humid conditions of places like Burgundy or the Champagne region of France. Some grapes want warmer climates that may also be drier—these would include the grapes of the northern Rhône such as the *viognier* and *syrah*. And some varieties thrive under the even hotter and drier conditions of Provence. These include the *grenache* (both the *noir* and *blanc* varieties), *mourvèdre*, and *syrah*—the *syrah* is a more flexible grape and is a big part of the wine culture in both of the Coteaux in the Coulon.

But the variety of grape is only the start of making good wines. After all, almost all grapes are made up of the same stuff— 75 percent pulp, 20 percent skins, and 5 percent seeds, with small traces of some minerals and other compounds thrown in. The factors that go into making that pulp-skin-seed mixture into decent wine depend on the skill of the vintner, the soil, and the sun. Some people, especially the French, refer to all the factors that produce a wine en masse as the "*terroir*" of the wine. *Terroir* includes a multitude of characteristics in the physical landscape. Many people argue that the bedrock and the soil are the most important factors—in the two French AOCs the soils are originally weathered from limestone terrains that are high in calcium carbonate. Others think climate is the major contributor to a grape's fate. The Coulon's hot, dry summers and cool, moist winters are perfect for some varieties, especially when the mistral

Typical vineyard below the small village of Goult in the Coulon Valley.

blows and cools the vines in July and August. Some think that in addition to the overall macro-climate, the micro-climate of specific sites is the critical component of *terroir*. Characteristics such as which direction the slope of the vineyard faces can be crucial—in the Northern Hemisphere, south-facing slopes will inevitably be consistently warmer than north-facing slopes. So *syrah* grapes, for example, might be best suited to the northern exposures in the Coulon while the *grenache* is more comfortable on south-facing ones. Other physical factors can complicate defining a specific *terroir*—for example, when does the rain fall (spring, summer, winter?), how deep are the soils, what micronutrients are available to the vines, and many more. *Terroir*, as you probably have guessed, is a very complex set of conditions that interact to produce a grape of a certain quality.

On top of this entire physical milieu, we need to also include what the vintner, or *vigneron,* does to the grapes before and after harvest. How are the vines pruned, how close together are they planted, and how drastically are the grape bunches thinned are all questions that deal with the *terroir.* I cannot imagine that a *vigneron,* when starting a vineyard, does so without examining the region's culture and lore either. Is there a tradition of manual labor in the area, or are most vineyard tasks done by mechanical means? Are oak barrels used for the aging process? Are the barrels new or old? Are the barrels from French oak or American oak? How long should the fermentation last? The aging? In a way, the work and complexity the *vigneron* must deal with remind me of the 1960s–1970s rock group Blood, Sweat, and Tears. All of these human factors, I would argue, are also part of the *terroir* of a wine. The end product is more than the sum of its parts because when these parts are put together, they create something bigger than themselves. The magic of wine is the alchemy of place and people that creates the corked liquid we enjoy so much.

Wine has been produced and drunk in France, especially in Provence, for thousands of years, yet about 150 years ago the entire wine industry in France and in Europe more broadly was in serious jeopardy. In the small but up-and-coming appellation of Lirac, just a few miles west of the Coulon, the drama began in 1863. A small *vigneron* imported some vines from the United States. No one knew that the American vines were carriers of a pathogen that the vines from the New World were resistant to but the European vines were not. The world of French wine was turned upside-down when the *phylloxera* aphid spread rapidly and destroyed at least three-quarters of the production in France and eventually in the rest of Europe as well. *Phylloxera* deforms the roots of the vines and carries a fungus that girdles the root and kills the plant. Obviously, the search for a solution to the little pest took on epic importance. The problem came from North America, and so did the solution. The problem was solved by grafting the noble *Vitis vinifera* grapevines that produce the high-quality grapes for most wines onto *Vitis labrusca* or some other North American grape species roots. The North American roots were carriers of the disease but were also resistant to it, so this procedure temporarily solved the problem. Here in the twenty-first century, however, the wine industry is not out of the *phylloxera* woods because slowly but surely the *phylloxera* is adapting to the North American rootstock, and these roots are starting to become vulnerable, as has already been seen in a few AOCs. A new solution to the problem will need to be devised in the next decade or two. This time at least it will not be the complete surprise the first invasion was—we know it is coming.

We can now shift to talking about wine in general and specifically about wine in the Coulon and the flanking uplands to the north and south. Most of the wines from the United States, including Colorado, and from places such as Australia and Chile are varietal wines made from a single grape variety or are produced in areas where one grape is overwhelmingly dominant. The wines of France, on the other hand, are almost all blends of two, three, or more varieties. While California and Colorado produce char-donnays or merlots or pinot noirs, France produces Bordeaux and Burgundies and Côtes du Rhônes—all named for the wine region from which they come and not for the grape or grapes used. The Châteauneuf-du-Pape AOC, for example, allows a mix of up to thirteen varieties of grapes in regulated quantities, creating a massive task for the wine maker in balancing tastes and maintaining quality. Unless you do a lot of detective work, you may never know what grapes went into your favorite St. Emillon. The rules and regulations for an AOC, for example, will include what grapes can go into a "red" wine or a "white" wine or a "rosé," what percentages of those wines are certain grapes, and even how the wine is made, but you will seldom see the proportions of grape varieties used listed on the bottle.

Very little exceptional white wine is produced in either of the Vaucluse AOCs. Most good to great whites are from grapes that grow in somewhat cooler climates, which definitely leaves out southern France. The whites produced here are decent but not noteworthy. The reds and the rosés that come from both AOCs are the signature wines of the Coulon. Many people know the gist of how wine is made, but I think it is important to talk about the general difference between a red wine and a rosé—both of which can, and often do, come from the same grape varieties.

The color and much of the flavor of red wines come not from the pulp where the juice is located but instead from the skins and, to a lesser extent, the stems and seeds that may or may not be left on during the initial fermentation of the wine. The deep red color associated with most red wines and the tannic bite in many of them come from the grape skins left in the "must" during fermentation. This first fermentation may extend from a few days to a few weeks. A rosé, in the most basic terms, is a wine in which the skins are left in the must for a much shorter period than they are for a red wine. This gives the wine the pinkish blush to deep pink colors wine drinkers associate with a rosé. Most French (especially Provençal) rosés are much more robust and less fruity than rosés we are used to in the United States. They have body and a lot of rich, deep flavor.

Both the Côtes du Ventoux and Côtes du Luberon are known, at least locally, for their reds and rosés. One of life's small pleasures is drinking a chilled Côtes du Ventoux rosé on a hot Provençal afternoon. As in many places, the reds and rosés are from the same blend of grapes, if not the same proportions. The mainstay of grape varieties, or *cépage,* in the two AOCs is the *grenache.* This was originally a Spanish variety, where the grape is called the *grenacha.* This grape produces high sugar content and therefore high alcohol content—up to 17 percent at times. It has a spicy kick that is very distinctive. Its downfall is that it does not produce the rich, deep red color people like to see in their red wines. To enhance the color and add to the spicy, peppery taste, the *syrah* grape is almost always added in goodly amounts. At least 60 percent *grenache-syrah* mix is required under the rules of both of these AOCs. *Les vignerons* blend the wines each year, or vintage, to create either the consistency they want from year to year or the exceptional vintage that comes along only once in a while. Added to the *grenache-syrah* blend in most of the wines are various smaller amounts of *cinsault, mourvèdre,* or *carignan* to add a softer touch, a tannic edge, some rich red color, or more intense body, depending on the blend desired. A few other minor grapes may be a part of the final wine, but these main characters are found in almost all the reds and rosés from the two AOCs.

The people of the two Coteaux are justifiably proud of their wines, even if they are less famous than their neighbors to the west in the Rhône Valley. Most of *les vignes* are small, from ten to fifty acres. Dozens of examples of their wines can be found in the Maison de la Truffe et du Vin du Luberon in Ménerbes, where rack upon rack of Côtes du Ventoux and Côtes du Luberon wines are displayed in the cellar of the beautifully restored building dedicated to wine and truffles. Even more fun than visiting the Maison in Ménerbes is going to the individual *caveaux*, where visitors can see the wine as it ages and taste the wine produced from the ground on which they stand.

The Maison de la Truffe et du Vin du Luberon was started by a former mayor of Ménerbes, the small hill village on the southern edge of the valley. The former mayor is a *vigneron* of a well-established and respected winery about two miles west of Ménerbes. The Domaine de la Citadelle is a beautiful yet unpretentious complex of winery and cave that also hosts the Musée du Tire-Bouchon, or Corkscrew Museum. The complex could be the symbol of small wineries in southern France, as the limestone walls of the buildings blend into the surrounding vineyards. You may have tasted some of their wine even if you have not traveled to France or Provence because this is one of the select few wineries from the valley that exports to the United States. I was pleasantly surprised when I found that a wine I had tasted in Colorado came from the winery I was visiting in the Coulon Valley.

The Citadelle's production of wine is a great example of the idea of the *terroir* in French wine making. They have very specific vineyards in very different places for the different quality and price levels of the wines they produce. For example, they make a nice *vin de pays* that is oddly made from 100 percent cabernet sauvignon—a noble grape that makes great wines in other regions of France and the world but is relatively mediocre here. The cabernet grapes are grown in the lowest-elevation vineyard the Citadelle owns on alluvial soils near the farming community of Coustellet in the valley bottom. Other crops, such as cherries, are grown in this fertile soil, and yields for the cherries and the grapes are high. The vintners get substantial harvests from the vines, which do not produce a great deal of character. Therefore, the wine from these grapes is good and plentiful but not special.

The other three red wines the Citadelle produces move up the slopes—the higher and harsher the soil, the better the wines and the more stress put on the *vigneron*. But the stress pays off, as predicted by the Provençal saying at the beginning of the chapter—"Make me poor, I will make you rich." The proverb means the Citadelle's Le Châtaignier wine, which comes from the vineyards surrounding the winery and cave compound, is good and very consistent, but will probably never be great.

The next level in both elevation and quality of the wine is Les Artèmes—a 40 percent *syrah*, 40 percent *grenache*, and about 20 percent *mourvèdre-carignan* blend. This wine has much more character and can be kept for up to ten years. The vines from which this wine comes grow on soil that is somewhat less fertile than the soils at lower elevations, and the vines are older—at least twenty years. Production is limited to 40 hectoliters per hectare (about 390 gallons per acre). This limit is accomplished by literally cutting off (green harvesting) developing grape bunches to decrease the yield and allow the vine's resources of nutrients and water to be used by the remaining grapes left on the vine.

The star of the Citadelle wines is the Gouverneur St. Auban, grown on some of the winery's highest and harshest ground— the Hauts Artèmes Plateau. This high land has much less fertile soil, has more vagaries of weather, and builds significant character into the grapes. The St. Auban is a *syrah*-and-*grenache* blend from even older vines, and production is limited to 30 hecto-liters per hectare (about 300 gallons per acre). It is a wine that is aged longer than others and in oak barrels. A taste comparison of these wines reveals an amazing progression of flavor as the vines get older. It is as though the roots of the vines soak up the essence of rock and soil as they age. If a wine lover wants an intense, earthy wine, choose one that comes from older vines.

As a counterpoint to the well-heeled, well-established, internationally known Citadelle, we visited a small winery near the village of Joucas across the valley from Ménerbes. The Domaine de l'Auvières is one of the few organic, or *biologique certifié,* wine producers in the region. We would call it a mom-and-pop winery in the United States. It looked as though the tasting room were an extension of the small stone house, and the entire enterprise is so unprepossessing that it is easy to miss if not for the small sign on the roadside. There are no picturesque vineyards surrounding the house; in fact the house and the cave sit in the scrubby landscape of the *garrigue*. Their ten hectares, or about twenty-five acres, of vines are located elsewhere in the valley.

We tasted and bought two wines that were surprisingly good and cheap. One was a *syrah-grenache-cinsault* blend that was not aged in oak; the other was the same wine aged in oak. Both were substantial wines with a complexity one would not

Example of "old" vines near Roussillon. The grapes from these vines make wine with a deep, earthy taste that is usually more expensive than wine from younger vines.

expect from something in this price range. The Domaine's vines produce about 32 hectoliters (320 gallons) per acre, equivalent to the production of the high-end wines of the Citadelle at about a third the price. A lesson learned, but probably soon forgotten, is that one cannot tell the quality of a wine by the looks of the winery or the price of the wine. The best and probably the only way to distinguish among the various *vignerons* and their wines is to taste them—not an onerous duty.

The wine regions of the Coulon may get only a quarter-page in the wine encyclopedias, but the wines of the North Fork seldom rate even a footnote. They are not only secrets but more like top-secrets of the enology set. North Fork wines are just now being noticed in Colorado, where most of the wine production has historically centered on Palisade and Clifton, about sixty miles west of the North Fork Valley. In fact Colorado as a whole barely rates a passing mention by most wine experts. This lack of reputation is interesting because the North Fork wines are fast becoming Colorado's premier wines in tastings and judgings. Although wine production in the valley is still very small and the industry there is in its early adolescence (or maybe even pre-pubescence), the collective culture of the vintners there is very focused and intensive, and the methods used constitute the best mix of wine science and art.

The land in the valley could be considered the quintessential prototype for vineyards. The mesas upon which most vineyards sit have a loose soil structure of cobbles and gravel that gives great drainage and has the needed nutrients at depth for the deeply rooted vines to reach—vines around the world love this combination. The alluvial wash from the Grand Mesa and the remnants of weathered basalt provide a nearly perfect foundation for a wine enterprise. The main nemesis, of course, is the climate. At its best, there is no better place for many grape varieties; at its worst, it can destroy a crop or quickly damage or kill the vines. The main culprit is temperature extremes, at least at the low end of the thermometer. Yvon Gros, vintner and innkeeper, said that in December 2006 the temperature went from 60° F to 10° F in less than twenty-four hours. If this had occurred after bud break in the spring (a distinct possibility), it would have been the end of the grape crop for that year and maybe even killed off the vines. As it was, the December temperature drop merely lowered the subsequent fall harvest by about 30 percent by killing off some of the primary growth of the stem tips. One of the many things Yvon and other growers in the valley do to preclude total losses is choose grape varieties developed for cooler to colder climes.

The three most productive grape varieties in the area are merlot, cabernet sauvignon, and chardonnay. All three are also grown in such disparate regions as New York, Bordeaux, Chile, and Washington State. They have a wider range of temperature comfort than many other grapes and can withstand the sometimes very cold winters in the North Fork. Yvon uses some chardonnay that other growers in the valley produce, but he has taken a big leap of faith into other varieties. One of the two most important grapes in his vineyard is the Cayuga, a hybrid bred for the cold slopes above the Finger Lakes in upstate New York. He makes a white wine from the Cayuga that could be compared to a chardonnay without the oakiness most American producers create. His second grape is the relatively obscure chambourcin, a merlot–cabernet sauvignon hybrid. This is an intensely purple-colored grape that ripens late. The chambourcin is resistant to a number of fungal diseases, is very high in tannin, and produces a seriously full-bodied red wine.

One afternoon while Carole and I were visiting the inn, Yvon suggested gently that he and I check on the oak barrels aging in his winery (i.e., his big double garage). How could I resist such a suggestion? We started at one end of the aging line of barrels as Yvon extracted small glass-sized samples of wine from each barrel with a "wine thief." Yvon handles the wine thief, a large pipette-like apparatus used to get small amounts of wine from a barrel, like the expert he is, although he often gets too much wine in the thief—more of a grand larceny. We had taste after taste of his variously aging stock. The younger chambourcins were so tannic they puckered the lips. As we arrived at the more aged barrels, however, the wine became a delightful, robust, earthy red wine—just how I like it. I had never thought much about chambourcin—in fact Yvon was the first vintner I knew who uses it—but he seems to be on to something. At the end of just "a little" sample from each barrel, I was thinking how great it all was.

Most of the vintners in the North Fork blend the best of the latest science of wine making with organic growing practices and Old World wine making. They don't seem to be in the business just for the business; instead, they think about what is best for the wine and the vines and how to get the best vintage. As an example, I asked Yvon if he thinned or green harvested his grapes—a time-honored, often AOC-required way to concentrate the sugars in the remaining bunches. Yvon takes each year as it comes and carefully determines if thinning is the right thing to do year by year and variety by variety. Following the December cold plunge I mentioned earlier, the harvest was already diminished by 30 percent. But in more normal years he would have green harvested when the grape bunches started to set. Then at véraison—just as the grapes start to ripen—he may or may not have thinned to get rid of lagging bunches that would not ripen fast enough for harvest anyway. He also does not use chaptalization, or adding enhancing sugars to his wine. This is a trick some wineries use when they do not thin and the grapes have lower sugar values. Chaptalization increases the sugar and therefore increases the alcohol levels. Yvon uses some chaptalization in specialized wines meant to be "fortified"—for example, when he makes his port from the chambourcin.

Yvon is a sophisticated yet jovial French farmer who knows everything and everyone in the valley. With his talent for engaging all, he guided me around the valley to meet other vintners. This was a real education. Our first stop was with Len, a regionally famous artist who is good enough to make a nice living from his art. But maybe the real love of Len's life is wine and viticulture. Len has just an acre of vines, but he uses it as a sort of test plot for grapes. He has planted four or five grape varieties from which he makes wine—if after a few years he is not happy with the quality of the grapes or wine he gets from one of his varieties, he pulls up the offending vines and plants another test grape. He does not need to make a living from his land; he wants to see what *cépage* will succeed in making a better wine. Len characterizes the culture of the vintners of the valley—they are passionate, smart, risk takers, and scientific all in one. Other aspects of Len's place make the North Fork and the Coulon come to mind. His house and land could be transported to France and no one could tell that they did not originally come from Provence. His house is typically Provençal and is surrounded by his grapes and a menagerie of other fruit, including apples, peaches, plums, apricots, and artichokes—many of the same crops found in the French countryside.

Another of Yvon's colleagues, friends, competitors, and fellow vintners is Mike, whose vineyard is in the valley bottom—unlike most of the others, which sit on the mesas above the river. Unlike Yvon and Len and many other growers in the valley, Mike does not grow organic grapes. He uses little fertilizer or pesticide, but he does use some. If Len and Yvon are scientific about their grape growing, Mike is the Nobel laureate of the group. He analyzes every minute detail of his vineyard. We spent about an hour with Mike, and the topics ranged from acid levels and pruning strategies to brix values—the total dissolved compounds in the juice of the grape. Since most of these compounds are sugars, brix is used as the surrogate to denote sugar levels and thus potential alcohol levels. As we were standing in the midst of his vines discussing these arcane topics, I was wondering what his wine tasted like. Much to my dismay, no samples were offered.

In the nineteenth and early twentieth centuries, western Colorado had a booming grape and wine industry. The Grand Valley was awash in vineyards. Then in 1916 the state of Colorado imposed a statewide prohibition on alcohol, followed four years later by the Eighteenth Amendment to the US Constitution. The wine industry in the valley and nearly everywhere in the United States was instantly decapitated. When prohibition ended in 1933, virtually no vineyards were left save a few that grew table grapes. It took until 1968 to restore even a semblance of commercial wine production in Colorado. That year Gerald

Ivancie began the first new winery in the state, and even then he had to use California grapes because there were not enough Colorado grapes to produce more than a few cases of wine. Since then, however, the US wine world has changed, with huge production in California, Washington, and Oregon. Other states, including Colorado, are starting to contribute to the overall US levels of wine production, although they produce much smaller quantities.

An indication that wine is becoming mainstream is the creation of a Colorado State University extension and experimental station devoted to viticulture near Grand Junction. The Orchard Mesa Research Center conducts intensive studies about the best grape varieties, pathogens that are a concern and how to deal with them, the water needs of vineyards in western Colorado, and many statistics for seasoned growers and those who want to take the plunge into viticulture. One look at the table of contents of the research center's website vividly illustrates the official state interest in wine production. The topics studied and advice to vintners include:

+ Grape Pest Management Workshop Materials
+ Colorado Grape Statistics
+ Grower Survey Forms
+ Grape Varieties / Rootstocks
+ Grape Cultural Practices
+ Resource Material
+ Grape Pests
+ Grape Diseases
+ Cold Hardiness
+ Grape Cold Hardiness—Orchard Mesa
+ Grape Cold Hardiness—Rogers Mesa

There are also sections on sustainable organic production (unheard of a few years ago) and a long list of publications. The center's scientific staff now includes viticulturists, enologists, soil scientists, and several other scientific subdisciplines dedicated to improving Colorado's wines and the Colorado economy.

Regionally, wine has become as important as other crops, especially orchards, on the western edge of Colorado. It helps that wine has such a high economic value. When Yvon and his friends experience great vintages, the extension service—which is funded by state tax dollars—sees money flowing into state coffers.

A sign that the United States is trying to catch up to international standards is the national scheme developed in 1983 that is similar to the French AOC system. The US system is usually much less stringent about how wines are made and what grapes can be used, but it does provide some quality assurance. The US system is called the American Viticultural Areas (AVAs). Colorado has only two recognized AVAs. The first is in the Grand Valley, which includes Palisade, and is appropriately called the Grand Valley AVA. The second is the West Elks AVA, which is really the North Fork Valley AVA but perhaps the name of the mountains just to the east of the valley gives the AVA more panache. The name may be appropriate because the West Elks AVA has the highest vineyard in the Northern Hemisphere. This elevated grower is the Terror Creek Vineyard and Winery on Garvin Mesa just east and above Paonia. It sits about 6,350 feet above sea level—it would be the highest in the world if not for a vineyard in Argentina that is above 7,000 feet. The Terror Creek vineyards occupy about ten acres—small by Provençal standards. They produce several cool-weather grape varieties including chardonnay, gewürztraminer, and pinot noir.

Whether one is dealing with an AOC or an AVA, the goals for the final product are a high-quality, consistent, guaranteed, honestly labeled wine—in other words, a known quantity that can be counted on. No one can take into account the vagaries of weather, pest infestation, or other circumstances, but at least there is a modicum of assurance that what you pay for is what you get. Generally, the AOC regulations are much stricter than those of the AVA, and in some cases they are downright arcane, but both the AOC and the AVA were developed primarily by the *vignerons* themselves to ensure their own wine futures. They are also, in most cases, enforced by the regional wine industry. Because the rules are made and overseen by the wine producers, the likelihood that they are followed is very strong.

FOOD

T HE SHARING OF A MEAL with friends or family is
one of the most universal joys nearly all cultures pos-
sess. Something about a communal dinner, lunch, or
even breakfast often brings out the best in conversation, interesting discussion, laughter, and thought. This might
be because food is such an intimate thing—we humans literally
take it into our bodies as we take in few other things. Numbered
among these rare items are water, air, and the occasional glass
of wine. Maybe this is why the communal meals we have had in
both the North Fork Valley and the Coulon are so memorable
and almost always bring a smile to our faces.

The laughter part began early during one communal meal we had with Yvon, Joanna, and other guests at the Leroux Creek Inn. Yvon gets most of his food for these meals from the gaggle of organic farmers and ranchers in the area. This particular night Yvon was making roast pheasant, which came from the pheasant farm up on Redlands Mesa. Yvon claimed with a straight face that for a few days before the meal he had driven up on the mesa at very high speeds in hopes that he might get some roadkill of escaped pheasants. He swore that this was what we were eating that night. Whatever the pheasant's provenance, it was delicious. Even if Yvon had not spent hours aiming his truck at birds on the mesa, these meals are never simple affairs. They always start with a selection of aperitifs—calvados, white wine, sherry, or some other product made into alcohol. Conversation, aperitifs, and hors d'oeuvres can last a long time—an hour or two at least. It is interesting that the French seldom use their own term *hors d'oeuvres*—instead, they usually call appetizers *les entrées*.

Many of the people in these dinner groups may be strangers when the evening begins. But the free-flowing aperitifs and the joy of the evening's good food and conversation soon have people interacting like old friends. By the time the "roadkill" pheasant was served, everyone was hungry, happy, and convivial. The main course, or *plat* according to the French, can last another hour or so, and this one was no exception. The selected wine from Yvon's cellar flowed easily and often—luckily, the diners only needed to walk to their rooms. Dessert brought some moans of gluttony, although everyone seemed to be able to eat his or her share. And there is often dessert wine and cheese to top off the night.

One communal dinner we had near Ménerbes was very similar but without the roadkill. Yves and Françoise own a renovated *bastide* (a country house in the south of France) just a few kilometers north of Ménerbes—the Coulon is within easy walking distance from the inn. Their *bastide* has six rooms, and that night all were occupied, so we had a dozen people at the petite fête plus our two hosts. *Mais oui*, we started with the aperitif. Unfortunately, the selection of drinks included only a nondescript white and rosé wine and the Provençal drink *pastis*. I would have chosen the *pastis*, but this cloudy-looking concoction is a hot-weather drink, and the weather that May night was decidedly un-summer-like. But things looked up when the salmon salad entrée arrived. The same conversation (albeit in French), the same laughter, the same community of the North Fork Valley were evident—people were having a great time, and Yves kept everyone in stitches (at least if you could understand his rapid-fire French). The plat was a delicious roast lamb and potatoes, the best lamb I have ever had before or since with no exceptions. A good, local, and free-flowing Côtes du Luberon red wine helped the meal along. The dessert was a "remembrance of things past" for me, as Proust might say. It was a rich, dark chocolate mousse over an alcohol-infused cake. My mother-in-law made a cake like this every holiday, especially Christmas and Easter. She would put so much rum or cognac in it that the cake dripped from your fork. The cake Françoise made did not drip, but it evoked the memories nonetheless.

This particular evening ended with the four French pharmacists, who always vacationed together, inviting us to dinner with them the next night. Carole and I took this as a huge compliment because even though the French are gregarious and fun, they have a natural reservation that keeps them from becoming too friendly too fast. It probably helped that one of the pharmacists had spent a great deal of time in the United

States and picked up some of the American casualness about social events.

This concept of large, communal, excellent dinners—often called "farm dinners" in the North Fork Valley—is catching on more quickly in the United States, particularly in the West, than it is in France. Chef-prepared, locally grown, fresh-food farm dinners have recently become an annual rite in Paonia and some of the other small towns around the state. There is a large group of these feasts in the wine regions of California, Washington, and Oregon as well. People love the combination of fresh local (often organic) food and good local wine. These dinners often take place in the best dining room of all—under the stars on a warm summer or early fall evening.

There is an uncanny similarity between the French and American versions of these dinners. From aperitif to dessert and cheese, they could be interchangeable but for the language. But this similarity does not extend to the rest of the day's meals. Breakfast, or *le petit déjeuner*, is an example. The French version of a refined breakfast is seemingly in a different universe from the North Fork version. The French breakfast can almost be encapsulated in a single word—bread. The French love their bread, and breakfast is usually a smorgasbord of varied breads with sweet, unsalted butter, local preserves or jams, and maybe some yogurt served with dark coffee so rich that it seems it could get up and walk across the room. When we stay in a place in France that does not serve breakfast, my morning routine is to get up early and walk to the local *boulangerie* for a freshly baked baguette or a *ficelle* or two. We will tear large chunks off the loaf (no bread knife here) and slather them with butter and jam. Washed down with coffee or tea, it is a hearty, if redundant, start to the day.

In contrast to the nearly unwavering French tradition of doughy *repas*, the normal breakfasts in most places in the North Fork Valley are almost always a unique creation that is often huge and varied. For example, Yvon fixes a breakfast of his own creation: a soft, hot flour tortilla stuffed with black beans, tomatoes, salsa, cheese, and seemingly whatever happens to be lying around the kitchen—all topped with a couple of poached eggs. The meal starts with loads of coffee and juice and freshly baked muffins stuffed with in-season fruit that is locally available. This elaborate concoction is so filling that we probably do not need lunch—although we always eat it anyway. The mix of southwestern flavors, combined with Yvon's touch as a chef, makes this one of my favorite breakfasts anywhere. The next morning he will serve something completely different; no two consecutive breakfasts are the same.

Although the French are famous for their cuisine and love of food, it is nearly always a letdown to eat lunch at any café that lines a street or sits on a plaza. For some reason I have yet to fathom, lunches often resemble the fare at many chain restaurants or a local fish and chips joint. Unless you go to an actual restaurant, often only one listed in the Michelin Guide or something similar, you will get only average fare at best. We never go to any of the "star" restaurants in the Michelin Guide. They are all very expensive and, although always very good, in our limited experience they are often too formal and serve too much food for our taste. But other categories in the guide rarely disappoint. These are the restaurants with one or two "forks"—the "*Restaurants agréable.*" They are usually much less expensive, much less pretentious, and have wonderful food and wine.

During one visit we were staying in the beautiful village of Lourmarin, and one Wednesday we decided to check out a particular "forked" restaurant in a neighboring village. We had been walking to lunch each day to someplace or other in Lourmarin, but we had heard about this great lunch place in Cucuron about 6 kilometers east of Lourmarin. We usually check the guide for the times and days places are open, but it never occurred to us that restaurants would be closed on a Wednesday. Well, it was not open that particular Wednesday, nor is it open any Wednesday, which we found out when the door was closed and locked and we saw the hours posted. We were disappointed but knew it was our own fault for not checking the Michelin Guide. We looked around Cucuron for another suitable place—there were only the usual sketchy sidewalk cafés with the usual marginal fare. We decided to try the even smaller village of Ansouis, another one of those villages that have been proclaimed "one of the most beautiful villages in France." Even though there were no listings in the guide, surely such a quaint and touristy little town would have a good restaurant open for lunch. We wandered the town's two streets but found nothing open.

We lowered our sights somewhat and went to a less famous village called Villelaure—nothing in the guide, and no open restaurants. The same was true in Vaugines, although there is a hotel listed in the Michelin Guide for this miniscule collection of houses but no restaurant. We then checked on Cadenet—there was a one-"forker," but it is closed on Wednesdays. Even though these villages are only a few kilometers apart so we were not driving far, it was still getting late, even for French lunches, and we were hungry after our simple baguette breakfast. We finally gave up and went back to Lourmarin. None of the forked restaurants was open on Wednesdays, but we knew of a *bistrot* that was open. It looked better than most cafés on the street, but after lunch there, we again swore that we would never again simply drop in to these kinds of places. We also learned what was so peculiar about Wednesdays. The schools in the region are all closed on Wednesdays, and the owners and workers in the better restaurants that are not economically challenged needed to be home when their children were not in school.

We dubbed this little adventure the "lunch from hell." A few days later we tried the Cucuron forked restaurant again for lunch and, checking the guide to be sure it was open, had the "lunch from heaven." Neither of us knows exactly what we ate, but it was one of the best meals we have had in France. The main course was some sort of roasted bird in a very tasty sauce. Both the service and dessert were impeccable, and it made us wonder why a place like this did not rate a Michelin star. I think we are among a multitude of people in France who wonder how Michelin decides on the stars. The mystery, however, is starting to be revealed, if only in very small doses. John Colapinto wrote a recent *New Yorker* article on the Michelin system of evaluation. He ate with one of the select Michelin reviewers in a New York restaurant and witnessed some of the machinations each review goes through. The reviewer's identity remains hidden, something that might not happen in France.

We grew to depend on the forks when we were not having a communal dinner, and the Michelin Guide became our culinary bible when we had to fend for ourselves. Some of these places were better than others, but they were all very good and worth the money. Our favorite for dinner was a tiny restaurant with ten tables or so in the village of Goult. La Bartavelle had two forks and was run by a married couple—the husband was the chef and the wife was the maître d' and waitress. When you made a reservation, the table was yours for the entire evening because the meal was expected to last at least two and a half hours but more likely three. Since the restaurant did not open until 7:30 p.m., the meals always ended late, at least by US standards. You had your choice of three entrées, four plats, and three desserts or cheese. Very simple menu, *non*? But even for my French-fluent wife, the descriptions of the meals were a real challenge. For example, a typical *entrée* at La Bartavelle was *Asperges de Goult en salade de jeunes pousses, frisure d'oeuf, speck croustillant et pomade d'herbes potagères.* This is only one item on the menu. We figured out that it was an asparagus salad with young shoots of something, something about eggs, and a crusty thing with ointment and potted herbs. Even our large Harrap's French-English dictionary did not include all these words, and it makes little sense even in English. But we trusted that the chef would serve something delectable and happily ate our slightly mysterious salad, ointment and all, with a sense of adventure and a satisfied grin.

When I think of all the exceptional meals I have enjoyed, both in Provence and the North Fork Valley, I have a vision of the paths the food in those meals has taken. In the next chapter I discuss the organic food movement in the North Fork Valley, for example. The valley is certainly not the only place where local, organic food is making an appearance on many tables in the United States and elsewhere. But the notion of healthy, tasty, environmentally produced food on everyone's plate is a grand scheme that is still a dream for most. Dining in the North

LE FROMAGE OU LE DESSERT

Le choix de fromages de chèvre de la ferme de Babeth et Cathy
***

Macaron "maison" à la menthe, chocolat mi-amer et fleurette au café
***

Mousse légère aux griottes, cerises éclatées
et crème glacée à la vanille Bourbon
***

ur un sablé, compotée de rhubarbe et fraises de pays, sorbet romarin

******

LE MENU A 39.- e

(prix net)

An example of the complex menus at good restaurants in Provence. This is a dessert and cheese (fromage) menu from La Bartavelle.

Fork Valley and in Provence may be an exception, as in these two places, wherever one looks or eats one finds just this kind of food. What can be more reassuring than to see the chef from the restaurant in which we ate last night buying the freshest local produce available in the outdoor market the next day for that

The Bartavelle sign with the namesake bird, similar to a quail.

evening's menu? In Provence and the North Fork Valley, this is a regular part of one's shopping experience.

Weekly farmers' markets have become the norm in many American communities. The crush of shoppers competing for fresh eggs or the best greens or succulent small carrots that do not have a shred of the usual cardboard taste is a well-choreographed contact sport today. The American markets are expanding in place and scope even as I write this. Our community, for example, has one market where only organically and locally grown produce is allowed, from grass-fed pork to unpasteurized, unhomogenized milk to the best vegetables—heirloom tomatoes, okra, garlic, beets, greens, beans—the list goes on and on. The buyers are eager, and the producers are passionate.

But the American markets cannot come close to the size, variety, and especially the longevity of the French markets. French weekly markets started in the distant past and are embedded in the French character—there may even be a gene in the French DNA that causes residents to shop at markets every week. Every city, town, village, and hamlet has at least one market; they have almost always been open one day a week in the same place in the town for the last 100 or 200 years. These markets can be huge and cover acre upon acre in the *centre ville*. In very tiny settlements they may be as small as one or two vans with the backs open selling a few vegetables or some meat that has just been butchered.

In the Vaucluse-Luberon area there are actually two very specific kinds of markets—*les marchés hebdomadaires* (the weekly markets) and *les marchés paysans* (the very local farmers' markets where only local, often organic, food is allowed). The *paysan* markets more closely resemble the markets most of us are used to in the United States. They are small, they are not found in every town, they sell locally grown produce, and they are very seasonal, open usually from May or June to November.

The larger, weekly French markets we have sampled are a wonder for the variety of things one can buy right on the street. One market day I tried to write down all the different stalls that were available. I created the long list below, but I got caught up in the frenzy of the scene and soon lost interest in writing everything down. The list I did compile included

* every conceivable vegetable known to Western civilization (plus a few others)
* *saucisson*—dozens of different sausages
* fish, mostly fresh and whole
* olives—do you know how many types of olives there are?
* olive oil—do you know how many varieties of olive oil there are?
* cheeses—same question for cheeses
* herbs—an entire stall of forty or fifty different herbs, mostly dried
* honey from lavender, clover, spring flowers, summer flowers, and on and on
* live goats and rabbits
* rotisserie chickens—these may be the only chickens we saw available to eat in all of our French travels
* clothes of all types—usually a stall just for traditional Provençal clothing
* olive wood products—bowls, cutting boards, salad utensils, and similar items
* jewelry
* beads—I have never seen so many varieties of beads
* dishes
* handbags
* umbrellas—usually more on rainy days
* sunglasses
* watches
* soap
* toys
* brass gongs—not sure of the French connection here
* two different makes of pot scrubbers, each in its own stall

A menagerie of dozens of olive varieties at a weekly village market in the Vaucluse.

Regarding this last item, we had passed by the stalls while they were setting up and scoffed at their silly hopes of getting anyone to buy an item that seemed as though it should be advertised on late-night television or some boring infomercial. We came back about ten minutes later, and people were packed six or seven deep awaiting a demonstration and the opportunity to buy these things that looked like colorful dead squirrels.

Where do all the people who buy items at these markets come from? Parking in French villages is always nearly impossible, but the markets take up whatever parking spaces exist. Of course many of the people of the village walk to the market—most villages are very compact and easily walked. But to attract the crowds we saw, there had to be more than just the locals shopping. There must have been a large contingent of people from the rural areas around the village and maybe people from other villages in the area as well.

When someone is called gullible, it is often considered pejorative. Gullibility is seen as a weakness of a naïve mind that lacks sophistication. This may well be so, but gullibility is sometimes openly used with no guile, just wise marketing. One evening, while we were visiting Forcalquier just east of the valley, is a case in point. My wife, son, daughter-in-law, and I made reservations at a local "forked" restaurant—arguably the best in town. That evening when we went into the village for dinner, we parked some distance away from the restaurant so we could enjoy a stroll through the busy streets and the festive atmosphere in the plaza. One particular restaurant on the plaza had a very brisk business, with many locals eating, drinking, and communing.

We noticed that the wait staff was somewhat overdressed but thought little of it.

Our restaurant was just around the corner from the plaza and up the alley, a little secretive and elite. We ordered a grand meal with great wine and a wonderful dessert. The bill was not insignificant but not out of line for a gourmet experience in France. When we left we went back around to the plaza and, just out of curiosity, went over to look at the posted menu for the "non-forked" restaurant we had seen earlier. As it turned out, the menu was exactly the same as the one from which we had ordered at the high-end place in the alley. We also noticed that the wait staffs in the two restaurants were either a collection of twins, or we had had the same ones in the establishment where we ate our gourmet meal. When we looked closely at the back of the plaza restaurant, we could see the seating for our restaurant through the long corridor. The only real differences were that the plaza was somewhat noisy and the prices were about half those in the alley. It slowly dawned on us that we *touristes américains* had not paid close enough attention to the lay of the land. It was a great joke on the relatively clueless visitors and a good lesson in being misled by appearances. We chose the ornate decor and hushed tones of the forked place when we could have had the less pretentious plaza experience for much less. Who could complain? We got what we wanted at a price we were willing to pay—*c'est la vie*.

Obviously, we had learned not to expect very good food at the sidewalk cafés of Provence. But we finally discovered a halfway house between them and the more expensive restaurants— the *salon de thé*. These small, cozy, usually women-run establishments are a combination of coffee/tea house, breakfast/lunch counter, and pastry café. They often have great lunches if you do not mind having little or no choice in what to eat; they usually have just one or maybe two choices on the menu. We often tried to find a *salon de thé* for lunch if we were not out hiking or seeking a big meal at a starred restaurant.

One day when we were visiting the ochre capital of the Vaucluse, Roussillon, we found a *salon de thé* owned by a German woman who was probably married to a Frenchman. We were the only guests except for one other couple. Since there were just the four of us and the owner, and the other couple was a Dutchman and a German woman, the conversation between tables soon began *à la franglaisdeutschdutch*. It was an enchanting conversation about our joint love of the Vaucluse and the Coulon. The gist of most of the conversation was unmemorable except for one statement made by the Dutchman. He said, in fairly good English, that he always took at least a case of Côtes du Luberon wine when they returned home at the end of the summer. But the wine never tasted as good as it did when it was drunk on a sun-dappled terrace in southern France on a hot summer day. I think that statement sums up the way a geographer might describe the influence of place on how we perceive our situation. The location, the weather, the season, the food, and the wine are all part of the place and contribute to how we feel and how the taste of the food and drink is experienced. Take one element out of the equation, and our perception of place is just not the same.

The North Fork has a much less established list of restaurants—it is, after all, still in its infancy as far as tourism goes. But there are a few consistently good places to eat. One such establishment is the Flying Fork in Paonia. It has an Italian flavor but is not really an Italian restaurant in the normal sense. It has all types of dishes, most made from local, organic farm fare. The Flying Fork is also a social center in Paonia and the valley. Any meal there will be accompanied by the conversations of locals who use it as a gathering place. There is also a Greek restaurant chain of two establishments—one in Paonia and the other in a renovated house on the main street of Hotchkiss. I expect that as more outsiders move to the valley, the list of good places to eat will expand rapidly.

The wonderful farmer-writer Wendell Berry often speaks and writes about the place food should hold in a society. He says, "Eating with the fullest pleasure—pleasure, that is, that does not depend on ignorance—is perhaps the profoundest enactment of our connection with the world. In this pleasure we experience and celebrate our dependence and our gratitude, for we are living from mystery, from creatures we did not make and powers we cannot comprehend." In many ways, I think Berry's reverence for food sums up the unarticulated love of eating and respect for food that exist for many in the two valleys discussed in this book. Eating, and the production of food that is necessary for it, can be a mystical, transforming, and cultural experience. Not all Americans, or I daresay French, agree with this. But food and our relationship to it can help us recapture our groundedness to the land and our place in that land.

Food and the activities that surround it—the growing of it, the selling of it, the cooking of it, and the eating of it—are among the most important and often enjoyable things we do in life. All too often we take that entire process for granted—many have heard stories of urban or suburban children (and sometimes adults) who think milk comes from the grocery store and that meat is just some commodity that comes wrapped in Styrofoam and cellophane. I am sanguine enough to hope that our two valleys, one in western Colorado and the other in southern France, are microcosms of a reawakening in our cultures of the importance of food and how we obtain it. I admit that it was a lot of fun to research this chapter, and I suggest a similar adventure for everyone who eats.

I include a final familial gastronomic anecdote for this food chapter. One summer several years ago our high school–age daughter traveled to France to meet her French relatives for the first time. This was, of course, cause for a large, outdoor family fête for the American cousin. All the aunts, all the uncles, all the cousins from near and distant gathered in Lyon for a celebratory meal that started in mid-afternoon. Our daughter was used to family gatherings in the United States with her French grandmother and southern grandfather that lasted an hour or two—much longer than the average American meal.

As the first dishes, along with wine, came out, our daughter—who was well trained by her grandmother—cleared her plate with relish. Then the next courses started to arrive, then in another hour or so the next courses came, along with more wine. After about two and a half hours our daughter was totally sated and with shock and a little awe realized that the main course had not been served yet, not to mention dessert and dessert wine and the cheese plate. She told the gathered throng that she had never seen so much food and so many courses in a meal. Her aunt, the hostess, replied *"le repas sans fin"*—the meal without end. This was an intentional play on words because it could just as easily have been the identical-sounding French phrase *"le repas sans faim"*—the meal without hunger. Since that summer trip, these two French phrases have become a family joke anytime we get together to eat and drink in the French mode.

SIGNATURES

Hot lavender, mints, savory, marjoram;
The marigold, that goes to bed wi' th' sun,
And with him rises weeping; these are flow'rs
Of middle summer, and I think they are given
To men of middle age.
—WILLIAM SHAKESPEARE, *THE WINTER'S TALE*

We are in the redemption business: healing the land,
healing the food, healing the economy, and healing the culture.
—JOE SALATIN, WEBSITE FOR POLYFACE FARMS

MUCH OF THIS BOOK looks at the many similarities that exist between the two valleys—nearly the same climate, nearly the same landscape, nearly the same cultural, social, and economic commitment to a place. But each of these places, and really every place on earth, has its own idiosyncratic qualities. I see these qualities as "signatures," those things that are essentially unique to a person or a place. And like a person's signature, some are easy to read and others are nearly incomprehensible. Because the two places have many signatures, I have chosen one from each of the valleys—they were the easy ones from a possible long list others could create.

In the case of Provence in general and more precisely the Vaucluse, its signature is the distinctive aroma of everything lavender. Lavender is grown and used in many places around the world, from New Zealand and Australia to Japan and Morocco to England and Ireland. But if any one place can claim lavender as its own—and actually does claim it as its own—it is the little corner of Provence along the highlands above the Coulon. In the small town of Coustellet, a few kilometers southwest of Gorde, sits the Musée de la Lavande, a museum dedicated to everything lavender and to a narrow definition of what lavender is and is not. One might think the production of lavender and the essential oils from it is a rather straightforward enterprise, with everyone involved attuned to the plant and its nomenclature. That is definitely not the case, however.

I have many friends and acquaintances who are ecologists and botanists, and each of them claims with all scientific certainty that the taxonomy, or classification, of plants is much more precise if done using the Linnaean system of plant identification. This system, in their minds, is the be-all and end-all for scientific nomenclature. The Swedish botanist Carl von Linné, or Carolus Linnaeus, came up with the system in the mid-1700s. As his name(s) reveal, he liked Latin, which at the time was the language of the learned. Linnaeus's personality was such that the more learned Latin names were appealing to him. The idea of a hierarchy of classes, with the most specific being the genus and species everyone would agree upon rather than using the vernacular or common name, was a good one in theory. The problem, of course, is that all botanists around the world do not necessarily talk to each other or agree with each other's assessment of whether a plant is a new or different species or not. Add to this the complexity of nature and natural variation in plants, and sometimes a problem occurs. For example, does a species exist only when reproduction can take place between two like plants? If so, how do hybrids occur, both in nature and aided by human ingenuity? Such questions are still being asked and may be even more prevalent now, when so much work is being done with DNA and genomes.

Lavender could be the poster child for this situation. "True lavender," as defined by most French citizens, is *Lavandula vera* for some but *L. angustifolia* for others. And some people also call it *L. latifolia*, *L. officinalis*, *L. spica*, or *L. delphinensis*. It all seems to be the same plant. But botanists admit that there are between twenty-five and thirty species of lavender, so not even they can agree on the total number of species to name. The passionate people who work at the Musée de la Lavande insist that the only, good, real, best, true lavender is *L. angustifolia*—they relegate all others to second-class or even non-lavender status. But a hybrid of *L. angustifolia* and *L. latifolia* (yes, the same name as above but a different species) known as *lavandin* is taking over the lavender oil trade because producers obtain about ten times the amount of oil from *lavandin* as they do from true lavender. True lavender is also more difficult to grow, requires more care, grows at higher elevations where there are more vagaries of weather, and has shorter stems, so it is more difficult to harvest mechanically. Not much of a choice unless you want to be a purist, and the people at the Musée de la Lavande are decidedly purists.

On top of the scientific confusion over names is the bureaucratic morass over regulation. French lavender production is regulated by a series of French organizations. The first is the national scientific organization called the Conservatoire National des Plantes à Parfum, Médicinales, Aromatiques et Industrielles, the CNPMAI. The top regulator on the commercial side is the Office National Interprofessionales des Plants à Parfum, Aromatiques et Médicinales, the ONIPPAM. This is the umbrella organization for three lower levels of bureaucracy, each with its own many-letter acronym. The French seem to like long, very descriptive, arcane names for their organizations, sort of like the US military, but in this case it smells a lot better.

Prior to the twentieth century, the lavender industry was very small and local. It depended mostly upon the wild lavender plants that grew on the high slopes above the valley at elevations between 500 and 1,500 meters (1,640 and 4,920 feet). Wild lavender grows in some very harsh places—stony, infertile soil in full midsummer sun. In fact it grows very poorly in good soil. Lavender resembles grapevines in the sense that the best results occur when the plant experiences a good deal of stress. The people at the Musée in Coustellet say that true lavender can only grow from seed and can only be found in four French departments—the Vaucluse, Drôme, Haute-Alpes, and Alpes-de-Haute-Provence. Any plants called lavender outside of these southern French appellations are really *lavandin* and not true lavender. *Lavandin* is usually propagated using cuttings, not seed. These claims are basically unverifiable in the scientific literature and in the publications at the Musée. To most people the essential oils from lavender and *lavandin* look, feel, smell, and even taste the same. Nonetheless, the staff at the Musée swears that the medicinal properties of real lavender are much more powerful and widely usable than the oil from mere *lavandin*.

More than just about anything else, this book is about landscapes, and the landscapes of lavender (*lavandin?*) are some of the most distinctive, picturesque, and evocative of any I know. The requisite view of these iconic lands is one that looks toward long rows of perfectly sculpted lavender bushes upward to stone-clad villages looming from a hilltop above the fields. I have never met anyone who will admit that the layout of these scenes is intentional, but my guess is that there is at least some suggestion by local commercial establishments that lavender farmers create these almost too perfect pictures on the ground. What tourist from the strip mall–obsessed world of the American suburb can resist taking the requisite photograph and stopping at the local café for un apéritif, then just sitting and soaking up the smells and tastes of this quintessential Provençal experience?

"It's all your energy, it's all your time and finances molded into a dream that no one else can see but you" is a heartfelt quote from John Cooley of Rivendell Farm in the nearly extinct little town of Austin, Colorado, just down the road from Hotchkiss. John is an organic farmer on a small plot of land who grows potatoes, heritage tomatoes, and a few other organic crops and has been president of the Valley Organic Growers Association— VOGA (sadly, not quite up to French acronym standards). If anything defines the North Fork Valley, if anything can be called its signature, it is a growing commitment by many people like John to real, nourishing, healthy, nonindustrial food that is usually organic. VOGA is a loose organization of many of these kindred food lovers and includes fifty-seven individual farms, vineyards, bakeries, markets, and ranches—most in the Hotchkiss-Paonia area but others spread across nearly a third of the state from Ridgway in the southwest to Almont north of Gunnison.

Like so many others in the organic food movement in the valley, John's farm is miniscule by American standards (about 10 acres), relatively non-mechanical, and about as far from the industrial-agriculture cartel as one can imagine. He works the farm by himself with some intermittent help from kindred organophiles, so even with just 10 acres John has as much land as he can handle. In fact much of the 10 acres is not even planted in crops but either has trees or lies fallow. Everything about the farm is intensive, yet it resembles chaos when compared with the 1,000-acre fields of a pesticide-laden, homogeneous wheat farm in eastern Colorado. For example, John is building up the organic matter in the soil so that no industrially produced fertilizers of any type are needed. He scavenges this organic refuse from wherever he can—when people are trimming trees, he is there; when they have rotting fruit that has not sold, he is there; when the horses . . . well, he is there. Many in the larger sustainability movement say there is no such thing as waste; everything is food or fuel in the long run. John and his colleagues make that saying a reality on one small farm after another. The organic content of John's soil has gone from about 400 pounds per acre a decade ago to over 3,400 pounds per acre today. No one in this movement monocultures, either. John has a wide mix of potato varieties, heirloom tomatoes, broccoli, melons, chilies, and sweet corn—all on about 3 to 4 acres of his property. Pests might get to one crop or one variety of crop, but the rest is often free and clear. No pesticides are used, only natural defenses.

As with many of his fellow organic farmers in the valley, John fears the Monsantos of the world. Concepts such as genetically modified seed and synthetic pesticides and petroleum-based fer-

Hand-built cordwood barn at the Rivendell Farm in Austin, a few miles west of Hotchkiss and along the Bonifide Ditch Irrigation Canal.

tilizers are anathema to VOGA. Agribusiness would like nothing more than to control every aspect of the way farmers work and produce food, from planting to cultivation to harvest to market. But in a strange way, agribusiness is afraid of organizations such as VOGA too because it cannot control them directly, only through lobbying and restrictive laws and regulations it can get passed by state legislatures, the US Congress, and officially sanctioned regulators.

John's history on his farm goes back generations, from the time his ancestors homesteaded the land and, in 1881, helped dig the Bonifide Ditch Irrigation Canal, which draws water from the Gunnison River upstream. These are very old rights, and since the Bonifide Ditch has the first priority on Gunnison water, the farmers who own it are virtually guaranteed water even during the driest times. The Bonifide draws 800 cubic feet per second from the Gunnison, which is allocated up to a maximum of 500 shares. One share is about enough water for 5 acres during a normal year. Forty farmers have rights on the Bonifide and use up those 500 shares. Rivendell Farm is at the top of the ditch, with other share owners all the way down the canal to Delta about ten miles west. Shares with their priorities can be bought and sold just like any other commodity. John's shares and his priority are like liquid gold, even more precious than gold since little unallocated water remains to be found in the arid to semiarid West.

The people in VOGA grow much more than food, however—they grow a strong sense of community too. Everyone knows everyone and knows who grows, produces, or sells what, and how good the products are. The sense that they are all in this vague enterprise called organic farming and ranching together lends an almost family feel to the valley. As with families, there are squabbles and animosities as well as support and encouragement. John, for his part, depends a great deal on Pat up the road for help. His potatoes come in early; her herbs come in later in the season. They each help harvest, transport, and market the other's produce. The growers and merchants here have an unofficial grapevine that is the epitome of efficiency. Word gets out when organic debris is available to gather for use in composting, when someone needs help with planting or cultivating, when the harvest should be starting, and any number of other things. Yvon at the Leroux Creek Inn is a part of this—he is an important conduit since he is gregarious, loves to visit other growers, and gives of his time. Yvon, for example, has his "chicken and egg" guy with an odd, cluttered farm on Redlands Mesa; he has the pheasant farm where he gets the "wild" fowl for some of the gourmet meals he fixes at the inn; he is also a real estate agent who knows what land is available and what prices are appropriate.

Some people are joiners and some are not. Lance, who lives up the road from Yvon and Joanna's inn, is an example of one who is not. He is not an official member of VOGA and is not in the group's multicolored brochure, but he may be the most dedicated to VOGA's principles of any farmer in the valley. He has about 5 acres of land in the scrubby pine and juniper woodland on Rogers Mesa. Lance has cleared some of that land and put up a small shed or workshop or barn-type building. His house is a beautifully crafted one-room structure, maybe 12 feet by 12 feet, which has a bed and a huge picture window that looks out proudly on his little spread and the expansive valley below. That is the extent of his "infrastructure" except for things that directly affect his crops, such as waterlines and trellises. As sparse as his accommodations are, his small fields—plots, really—are wonders of precision, color, science, and hard work. Everything is grown organically, even though he cannot afford to become officially organic certified. He experiments constantly to see what works best, and he uses very little other than muscle power to work the land. Without a lot of equipment, he can farm only a few acres, but they are productive beyond belief. His lifestyle is the quintessential example of going back to nature. Few of us could or would care to live such an austere, semi-hermit kind of existence. But I am very happy that there are people like Lance who do.

Two other small farms (they are all small farms) Carole and I visited are members of VOGA—the Small Potatoes Farm and the Zephyros Farm and Garden. Each is run on just a few acres but has a huge variety of organic produce, flowers, and goats' milk. The Zephyros Farm makes much of its revenue from micro-greens sold to high-end restaurants in the Roaring Fork Valley over McClure Pass, where Aspen is located. The Small Potatoes Farm is well-known for wonderful organic garlic braids and a palette of different-colored peppers. Both farms make it financially, but just barely. Organic farming at this scale is lucrative in lifestyle but not in monetary rewards.

This whole VOGA concept does sound somewhat quixotic—one of John's quotes is "All we need is given to us." The farmers work incredibly hard to extract these givens, however. Most of these people are stubbornly self-sufficient and independent. Most do not make much of a monetary living—they could probably make more waitressing or even bussing tables at a restaurant. But they share a larger, often unarticulated goal apart from just making money. They, like Joe Salatin in the quote at the start of the chapter, are altruistic enough to believe they can grow healthy food, heal the earth at the same time, and change the culture of how we see our food and our land. It sounds like tilting at windmills and few others see the dream, but for these people the dream is a hardscrabble reality.

HIKING

———————

Walks. The body advances, while the
mind flutters around like a bird.
—JULES RENARD

I HAVE JUST BEEN LOOKING AT A HIKING MAP for the
Grand Mesa—the big, flat-topped, volcanic mountain
that defines the northern horizon of the North Fork
Valley. The map was produced by one of the world's
great geographic organizations, and I have nothing but respect
for all of their publications. But what struck me immediately
after having spent a month living and hiking in Provence is the
paternalistic tone the map's information section takes. The map's
extensive legend section includes the "10 Essentials" for being
prepared. Such things as bringing food, water, a map and com-
pass, sunglasses, and matches are on the list. Then there are six

suggested hints for planning ahead, such as preparing for emergencies and scheduling your trip during low-use times. On top of that, there is a list highlighting travel and camping details, a list on how to dispose of "waste" properly, a list telling you what not to take from the area, a list telling you how to minimize the impacts of campfires, and a list telling you how to respect other hikers. It is exhausting to memorize all the lists and their exhortations. I feel as though I need a list of lists to keep track of the lists I am supposed to adhere to. Whatever happened to the simple mantra "Take only pictures, leave only footprints?"

I know these are all good, heartfelt suggestions and that they are essential to protect the hiking trails and keep hikers safe. But I more greatly appreciate the way most of the French trails are "managed." During our time in the Coulon Valley, Carole and I found two main hiking guidebooks for the Luberon region. Each has a three- to four-page section on each trail telling how to navigate the route. French routes are much more intricate, with many short legs that often cross private property, so the way must be well-defined. Most Colorado trails are on public land and are relatively straightforward and well-marked. In the Luberon guides, the systems for pointing out needed information are similar. For example, one book has three classes for difficulty and length and four classes for how technical the hike may be, from an easy stroll for *petits randonneurs* (young hikers) to a hike that may require some technical climbing expertise and equipment. There may be a warning or two in the texts about some extreme conditions, but that is it. They seem to treat hikers as if they have some intelligence and can make rational decisions for themselves. Of course much of this laissez-faire attitude probably stems from the fact that liability lawsuits are a rarity in France. If one falls and twists a knee or even breaks an ankle, the French attitude is "deal with it."

This certainly does not mean that all the Luberon trails are described clearly. Carole and I have a family tradition of getting lost at least once per hike while walking in France. Obviously, we always eventually find our way after some careful and creative translations of the French directions. A hiker may also face some phrases in the guides, such as the terrain is "supple and pleasant" or the trail "prances up the mountainside"—however a trail prances.

There is a certain adventurous character to the French trails. In addition to the fact that French trail authorities do not warn people about every conceivable eventuality, many trails have some very serious, fun sections that a normal American would call "hairy" or even a little dangerous. We had tried the Gorges de Véroncle trail once before but never completed the hike. We came back to conquer the entire gorge from bottom to top. About an hour and a half into the hike, we came to a place where the guidebook said almost offhandedly that you could use the installed climbing aids or your own rope to rappel down sixty or so feet and continue the hike. The installed aids consisted of a steel cable bolted into the rock, which you could hang on to for the first fifteen vertical feet. Below the cable was a vertical steel bar for the next fifteen feet, then a rope for another fifteen feet, and finally a ladder for the last section to the base of the gorge where you could continue the walk. We caught up with a slower group that included a few acrophobic members. These twelve French walkers, all members of a hiking club, were in front of us when we arrived at the top of the wall, and it took them nearly thirty minutes to get down safely. Some were clearly petrified and had never done anything like this before. There were no signs or instructions warning of potential death or catastrophic injury, no indications that hikers were responsible for their own actions, no signing of liability waivers—just enthusiastic, if somewhat careful, scared people enjoying a good challenge and some fun trying to negotiate the interesting trail.

For another attempted hike up the Gorges de Régalon— there are a lot of gorges in the Luberon and Vaucluse—the guidebook implied that it might not be a good idea to hike the gorge during or after a rain. It had been raining off and on for two weeks, but the last day or two had been dry. We thought we could probably do the hike; after all, how bad could it be? The hike began through olive groves and a pine forest and was very pleasant. When we got to the actual entrance to the gorge, however, we encountered a foot-deep torrent roaring out along the path/streambed. This is a gorge where some of the passages are only 50 centimeters (about 20 inches) wide, with many pitches up steep rock steps. Even if we could have kept upright going up the cascades, the waterfalls over the rock steps would have been considerable. We opted to find a different hike that day.

On our way back to the car, we had an interesting encounter with two couples from one of the Scandinavian countries—our guess from their accent. They were on their way up the same gorge we had just tried, and we told them about the situation they would find. The two women immediately agreed that it was too wet and dangerous for them to continue, but the two men wanted to forge ahead. The men said they had waterproof boots, and they were sure they could make it. We never learned the result of this domestic disagreement, but we were fairly sure that they did not get very far up the gorge. At least we did not read about their demise in the next day's paper.

One of the steel cables anchored into the walls of many gorges to aid hikers in negotiating interesting terrain.

One advantage to hiking in the Luberon and to some extent in the Vaucluse mountains is that the mountains are long, narrow, and steep. Therefore, the valleys on either side are laid out like an embroidered wall covering in a museum. The intimate, ancient field systems make these lowlands look like bucolic Currier and Ives scenes with vivid and contrasting colors and patterns. The Grand Mesa and West Elks abutting the North Fork Valley are much more massive, and the lowlands are most often hidden from view. In Provence we took a variety of hikes up numerous trails where there were spectacular views of the valleys below.

One of these hikes was above Oppède-le-Vieux through the Vallon de Combres. This hike took us above the ancient village of Oppède (versus the new village of Oppède that lies just below the old one), which was a papal estate during the period of the Avignon schism. The old village was abandoned in the sixteenth century but was revitalized by an artists' colony during World War II. The climb up the very narrow valley is strenuous, and the steep walls and slippery trail above the cliffs are invigorating. The guidebook warned that one part of the trail slopes toward the abyss, that it is slippery when wet, and that the fall, if a hiker slips, is about 300 feet. Traversing this section was somewhat tense, but the views of the ancient town and the valley below were worth the effort. At the far end of the trail one emerges at the top of le Petit Luberon and into a forest of Atlas cedars introduced from North Africa.

Another of these challenging but scenic hikes took us up above the small, now upscale village of Les Taillades. The village was once known for the skilled stonecutters who used the local limestone conglomerate for much of the building done in

A view of Oppède-le-Vieux from the hills of le Petit Luberon above.

the region over many centuries. The village is also remembered fondly for the large clay quarry perched high above the town on the edge of le Petit Luberon. The children of Les Taillades gathered clay each year to make the *santons* famous in crèche displays each Christmas. As with most of the Vaucluse villages and hamlets, Les Taillades was very poor until just a few decades ago. Only with the advent of the tourist and second-home invasion over the last twenty to thirty years did the village become the prosperous place we see today.

The climb took us high above the very southern end of le Petit Luberon, with a good view of Cavaillon a few miles to the west. A geologic oddity, known as a *klippe*, can be seen in the upland on Cavaillon's west side. This is a piece of the Luberon upland that was transported there by tectonic activity a few million years ago, which set down this chunk of rock like a wayward child's wooden block.

Probably the most intimate landscape view during our numerous hikes was from above the Vallon de la Sénancole below the slopes of the *garrigue* about 3.3 kilometers (2 miles) northwest of Gordes. The trail is called the Circuit de Cancouple and traverses the ridge high above the valley floor. The twelfth-century Cistercian Abbaye de Sénanque was built in a tranquil setting in the small valley bottom. The guidebook refers to the abbey's architecture as "austere." The colonnade that runs along the cloister arcade was built with minimal decoration; the severe beauty of these stone arches was said to lead to less "sensual distraction." The abbey still functions as an abbey, but because busload after busload of tourists click away with their digital cameras, parts of the abbey are closed off for use only by the resident monks still intent on pursuing the spiritual life. Seen from the trail high above the valley, the abbey's typical *tau* cross form is very evident. A main difference between the Sénanque Abbey and the multitude of other abbeys throughout Europe is that the main liturgical end of most abbeys faces east—toward the rising

Cavaillon as seen from the very western end of le Petit Luberon. Note the hill beyond the city. It was separated and moved away from the Luberon upland millions of years ago by tectonic forces.

sun and the obvious religious metaphor. But at Sénanque this part of the church has to face north because of the narrowness and orientation of the valley—there was no room to turn the structure ninety degrees.

While on the hike that took us to the abbey overlooks, we kept hearing chainsaws from various directions as we walked through the dense oak growth of the *garrigue*. It was a Sunday morning, and we wondered what was going on. We found out later that this is an area of intense harvesting of scrubby oak trees for making charcoal. Most of the *charbonnières* do this as a side-line to their normal jobs, so weekends are most convenient for collecting the wood.

The Abbaye de Sénanque as seen from the path along the cliffs above the small valley.

One of the pleasant and interesting surprises about hiking in the Luberon region and the North Fork Valley is the similarity in the vegetation of the two places. I have discussed the parallel nature of the major ecosystems—the pines and oaks of the *garrigue* and the mesas. But the wildflowers and shrubs are equally related. A vibrantly colored blue flower caught our eye on the French hikes, and we realized that this flower is almost an exact replica of the blue flax found all over Colorado. There is also Colorado's fireweed, which gets its name from being a pioneer after a wildfire. It was once referred to by its Latin name, *epilobium angustifolium*—a species very akin to the *épilobe à feuilles de romain* that grows on the many cliff faces of the Provençal gorges. Both places have lupines, wild iris, delphiniums, and common juniper. The *ciste cotonneux* reminds us of the sticky geranium in Colorado. This is just a sample of a list that could go on and on, reinforcing the idea of how close these two places really are in certain ways.

But the best flower display by far in the Coulon Valley and the Luberon more closely resembles what one might see in the golden hills around the Central Valley of California. The field after field of red (not orange, as in California) poppies that lie among the orchards and vineyards scattered on the valley floor are almost too gorgeous to believe. From high up on the hills you see the nearly random patches of deep, vibrant red, like a kind of natural Jackson Pollock painting. The spreads of poppies often cover entire fields and are so dense that no other flowers are part of the scene.

Finally, one of the critical (maybe the most critical) parts of French hikes is the lunch. The meal is almost always simple, with bread (of course), maybe a piece of fruit, and often some local goat cheese. The most essential part, however, is a bottle of the local wine. This should not be a grand cru vintage that costs a day's salary but rather a local, inexpensive one because all wine tastes like a grand cru drunk out in the open, sitting on a convenient rock after two or three hours of strenuous hiking and climbing. Just remember the corkscrew; glasses are optional.

We love to go to the North Fork in the autumn. It just feels right to be there when the Gambel oak is in full chromatic splendor and the year's crops are being sold in all the local markets. It is also time for the grape harvest and the feeling of being a part of the rich rural life of the place. One recent October we made our pilgrimage to the valley with hopes of eating, drinking, and hiking in the invigorating fall weather.

On the drive to the western slope we started to encounter some clouds, fog, and a few flurries along the route just before we reached Crawford. As we approached Hotchkiss, the weather was turning nastier, but we had not planned on hiking until the next day anyway, so we had no real worries about taking to the trails—we would simply see what the next day would bring. We awoke the next morning to one of those sparkling, crisp, clear days that make Colorado Colorado. We decided to try the Crag Crest Trail up on Grand Mesa since we had heard it was a great walk with stunning views from on top. We drove up and over Redlands Mesa to Cedaredge and turned north on Colorado Highway 65 to climb up onto the mesa. As we ascended we started to see some new snow lying peacefully in the sun among the Gambel oak, piñon pine, and juniper trees.

Typical hiking lunch in Provence.

When we reached the top, the roads were wet and slick, with a full covering of snow on the ground. By the time we got to the parking area across the road from Island Lake, there was a foot of new snow on the ground. We are moderately intrepid hikers and we really had our hearts set on hiking that day, so we decided to slog through the snow and see how far we could get.

The top of Grand Mesa is capped by an ancient basalt (lava) flow Hawaiians call "aa." This is not the smooth, ropey-looking lava flows we sometimes see coming down the slopes of Kilauea; instead, it is chunky and blocky and very rough. The trail up Crag Crest is well-established and usually easy walking. Of course it is meant to be seen and kept to. When the trail was covered with fewer than twelve inches of snow, we could follow its gist but with difficulty staying on the actual path. The rock on either side was sharp, protruding, yet hidden. We soldiered on for about a mile but finally lost all semblance of the trail on a steep, treeless slope just below the crest of the crag. We reluctantly surrendered and retreated back along the snowy footprints we had just made. We saw no one else on the route until we got back to the parking lot, where a family with small children had just decided it was not a great day for a stroll.

One early spring visit to the North Fork and the Grand Mesa gave us a much saner experience with the snow on the mesa. This time we came armed with snowshoes and friends who would show us how to use them. We had cross-country skied in Colorado for decades but had never used snowshoes. The top of Grand Mesa is perhaps the world's best venue for the novice snowshoer. It has loads of great Colorado powder, is relatively flat terrain, and has miles of easily followed trails.

This first snowshoe expedition was not really planned very precisely. We found a good pull-off from the road and saw some snowshoe tracks leading into the forest, which we followed. In the summer this route is an old gravel road along Kannah Creek, but in the winter it made the perfect test site for our newfound snowshoeing legs. We "shoed" for about three hours, had a great snowbank lunch, and came back tired but content at having attempted a new kind of locomotion on snow.

Carole and I love to hike and cross-country ski, but my favorite translocation mode is biking. On another spring trip to the North Fork we brought our bikes and planned to do some serious pedaling in the area. A local annual publication in the valley is *Our Side of the Divide,* which includes a good map of bike routes that might be interesting.

On this occasion Yvon and Joanna at the inn told us that the Sandhill cranes were just starting their migration through the area and had spent the night in and around Fruit Growers Reservoir just south of Cedaredge. This kind of grapevine news spreads fast in the valley. We wanted to see the cranes but did not have enough time to bike to the reservoir before they left for places in Utah. We accepted Yvon and Joanna's offer to take us and our bikes across Redlands Mesa to the reservoir, planning to do our ride after the cranes had gone. We luckily timed our arrival perfectly, as the first 100 or so cranes began their circling of the reservoir to gain altitude and a wind advantage with a cacophony of loud, rattling "hkkkkkk"s, as *The Sibley Guide to Birds* describes their call. It was an inspiring sight as group after group of 50 to 100 cranes took flight and thousands of cranes eventually departed to the northwest and their next night's bivouac in Utah.

We began our bike trip back over Redlands Mesa as the last of the cranes were barely visible on the horizon. The road back up to the mesa was steep, with many hairpin turns, but it was just what we needed to get warmed up. We planned on making a very circuitous and long ride back over the mesa since the straight-line return was only seven or eight miles. Once on top, however, we encountered a very strong wind blowing out of the northeast with gusts well in excess of 40 miles per hour. This kind of wind is not unusual in the spring, but it usually comes from the northwest or the west. The headwind we rode into required increasing effort the longer we pedaled. We stopped along the way to catch our breath and saw a golden eagle perched on top of a wooden power pole, some pronghorn bedded down against the wind, and many clouds scudding rapidly across the sky.

These human-speed movements provide the best way to immerse oneself into a landscape. The slow, deliberate movement through a landscape these measured modes of travel provide helps us sense the land and people's impact on it more fully. Every small hill, placement of a stand of trees, geometry of an orchard or field, or design of a building or town adds a little more to our own geographic understanding—assuming that we actually want to understand the world we are moving through. These slow movements through a scene do not guarantee comprehension, but they do allow us the luxury of seeing a revealed landscape slowly enough to take it in more fully. I learn more about a place during a slow saunter than I do through a windshield when I am driving at a high speed. The human senses are able to catch more of the subtle detail and small accoutrements of the natural and human world.

LA CHEVILLE
(THE ANKLE) INCIDENT

———◦◦◦———

He who limps is still walking.

—STANISLAW J. LEC

ASTILLE DAY, that celebration of *"Liberté, Egalité, Fraternité."* Our special French July 14 dawned somewhat cooler and fresher than the hot, sultry days before. It was almost invigorating—well, as invigorating as the low nineties in intense sunshine can be. Up to this point we had only done a few hikes in the Provençal countryside, and hiking is one of our favorite ways to get exercise and to really see the landscapes that are generally hidden from even the smallest roads. The inn where we were staying is only about a kilometer from the scarp that defines the southern side of les Monts de Vaucluse. Just up the road is one of the locally famous hikes up the substantial gorge carved out of the ubiquitous limestone—a perfect setup for a half-day, vigorous hike into terra incognita and our first attempt at hiking les Gorges de Véroncle.

Les Gorges de Véroncle cut their way through hundreds of vertical feet of limestone and run for several miles north into the higher uplands of the Vaucluse. The main gorge looks unlike most valleys or canyons—it is more like a deep surgical cut into the rock. But this cut was made by millions of years of running water, not a scalpel. At places along the worn yet rugged path, the gorge bottom can be as much as 330 feet below the plateau above. This is not a great distance by Black Canyon of the Gunnison standards, but this gorge is narrow, and at times it seems that hikers could almost touch both sides of the rocky slot if they spread their arms. A series of old abandoned mills (*moulins*) within the gorge were built to take advantage of the rush of confined water down the steep grades during the wetter season and after the more enthusiastic thunderstorms during the summer.

The path we were on that Bastille Day was an excellent example of the rugged, semi-wild experience of French hiking discussed in earlier chapters. The trail was narrow and criss-crossed the little creek innumerable times. There was a lot of undergrowth vegetation, including a remarkable number of fig vines flourishing because it was shady and less evaporation occurred than was the case in the broader valley. We stopped to examine the remnants of one of the numerous mills, the Moulin Cabrier, and continued up the ever more confined and steep parts of the trail. There were at least two near-vertical sections where some steel ladders had been set. Going up the ladders is not difficult, but without them it would be hard to continue the hike. After an hour or so, we had passed the more challenging

Le Moulin Jean de Mare in the Gorges de Véroncle, site of the ankle saga.

Metal ladder used for ascending and descending during hikes in the gorges of the Vaucluse.

sections of the lower gorge and come out into a wider, flatter alcove-like section of the route. It was nearly flat on the clean rock, almost like a sidewalk in comparison with what we had been traversing. I started looking around at the massive stone walls above us on either side of the gorge instead of staring at my feet and where I was placing them. The path was nearly flat, but not quite. I did not see the one-inch rock lip under my feet that was parallel to the walk, and in an instant we both heard a loud crack that came from my left foot as I rolled my ankle on this meager little irregularity in the trail. Carole thought I had kicked one rock against another, but I assured her in fairly loud, profane terms that it was indeed my ankle that had made the cracking sound, and I was pretty sure it was broken. I had never broken a bone before, but there was little doubt in my mind that I had done just that.

The French celebrate Bastille Day with even more military exuberance than Americans celebrate July Fourth. There was no way anyone from the military or probably even the local gendarmerie was anywhere near here—they were all at parades and fêtes in the cities, towns, and villages a long way away from our little stone alleyway. I was not going to be heroically rescued by anyone official for some time. Not much choice but to try to walk out. We were about a mile and a half to two miles up the gorge and above those rock walls we had scaled with the help of the steel ladders. The trip out was going to be interesting.

Like most people I know, I do not like pain very much. But also like most people, if you have no choice but to endure pain in an emergency situation, you endure it. I kept my shoe on to help keep the ankle from swelling too much; that would come later with a vengeance. A little digression here. Remember the Chaco factory in Paonia, mentioned in an earlier chapter? Well, I was not wearing rugged, ankle-high hiking boots or even an actual shoe during this trek. I was wearing a very nice pair of Chaco sports sandals—rugged soles with nylon straps but no ankle support whatever. I do not blame Chaco one iota. I was the one who did not want to carry those heavy hiking boots in my baggage on the trip from the United States. Most assuredly, I no longer wear sandals when hiking. I now leave them for the beach.

Carole found some substantial-looking branches along the way to use as makeshift crutches, but they kept breaking and making the hike out even less pleasant than it already was. I now carry a high-tech walking stick with me on every hike. The only way down the rock walls and ladders was facing outward, away from the rock, and scooting down on my good leg and buttocks. This was not a dignified sight, but it was unavoidable. By the time I had passed the most exciting of these rock shelves, holes had been rubbed into the seat of my pants. The trip out took about twice as long as the trip in, but I thought we made great time under the circumstances. Even an encounter with a snake across the path did not slow us down too much. We were told by other hikers coming up the trail that the snake was not poisonous, but it was big—at least four feet long as it crossed just in front of us. Luckily it ignored us. It was just before noon when we reached the mouth of the gorge. After emerging from the gorge somewhat bedraggled, hot, dirty, and a little exhausted, Carole ran the kilometer or so to the inn for the car. For some strange reason, probably because of shock, I thought it was a good idea to keep moving, so she found me slowly limping along the little road (broken ankle and broken hiking stick in hand) about twenty minutes later.

Some of the very, very few French whom we met who were unpleasant were those who owned and ran the inn in which we were staying. We had made reservations sight unseen from the United States using the Internet, which showed lovely photos of the place, but we were sorry we had done so. When we checked in a few days earlier, they had acted as though we were inconveniencing them by staying at their inn, and we should have been warned. When we got back to the inn after the trek out of the gorge, they were very skeptical when we said we were afraid my ankle was broken, and it took much haggling just to get ice to put on it. In response to our inquiry about seeking medical care, they informed us with some disdain that it was Bastille Day, and all regional hospitals were closed except the one required to stay open in Cavaillon—about 33 kilometers (20 miles) to the east. We got their directions, given grudgingly, and I hobbled back to the car to face the French medical establishment.

The drive to Cavaillon was not particularly unpleasant. The pain in the ankle was tolerable (shock can be a great anesthetic), and we had much conversational fodder complaining about the rudeness of our innkeepers. We would spend that night at the inn but would move to a much smaller, more pleasant place the next day. It took us a while to find the hospital when we reached Cavaillon. From the looks of the place, it came right out of the early 1950s and had not been painted or maintained since then. Some construction was taking place, which presented more obstacles for my rapidly swelling ankle to conquer. But we soon found the small door to the emergency room and walked (or hobbled, in my case) in.

There were no big, automatic sliding doors, no flashing lights, no nice overhang to keep folks dry in the rain. All we got was a small, unpretentious, very low-key entrance to the hospital— not much bigger than the doors to most houses. But we were aware of the bureaucratic reputation of the French and entered with some trepidation. What financial hurdles were going to be thrown down in front of us? What long waits for France's infamous socialized medicine? What gauntlet of forms were we about to endure?

We had been smart enough to grab our passports before we left the inn. The young nurse or assistant behind the glass portal in the small room we entered seemed pleasant enough and was concerned about my ankle. She kindly asked for my passport so she could copy the front page. Once that was done, which took about fifteen seconds, I was met by someone with a wheelchair and taken to an examination room. We were sure that somehow, the bureaucratic nemesis was lurking. About two minutes later another nurse came in and looked at my ankle and said it was probably broken, but the doctor would be in to look at it and X-rays would be taken. She apologized for the wait—it had been a total of five minutes since we stood outside the ER door.

The young, smiling, kind doctor came in about five minutes later and apologized for the delay. He said the ankle was probably broken, but I needed X-rays. Only one X-ray technician was on duty on Bastille Day, and she was working on another, more serious emergency and would get to me next. We waited another ten minutes, then I was wheeled into the X-ray room. The X-ray technician apologized for the delay and quickly took the needed X-rays. I was wheeled out to the hallway—the room I had gone to first was needed, and could I please wait for the doctor to read the X-rays? Again, they apologized for the delay.

The doctor came up about three minutes later and said that yes, the ankle and foot were broken in two places. But the good news was that the bones had not been displaced, so the treatment was to ice the foot and put it into a walking cast, which the French call an *orthomarché*. He put a wrap on it with some ice, then told us we had to get the painkillers, the *orthomarché*, and the crutches at a *pharmacie* because the hospital could not carry those items according to French law. The young doctor assured me that my family physician back in the United States would re-X-ray the ankle and do a thorough exam when we returned home. He again apologized for the delays. As they wheeled me toward the door I thought, "Here it comes—the money, the bureaucracy, the forms." What we found instead was an open door, a fond *au revoir*, and an apology for everything having taken so long—it had been forty-five minutes from door to door. In the United States I would still have been checking into the ER while they were making sure my insurance was up-to-date.

The Cavaillon Hôpital was old and worn-down, with peeling linoleum and drab paint. But it was clean, efficient, friendly, and competent. We were not dazzled by the architecture, but we were by the people and maybe even by the French medical system. Certainly the system is not perfect—I am not sure there is such a thing as a perfect medical system. We did have to find the one *pharmacie* that was open on Bastille Day to get the painkillers, crutches, and *orthomarché*. When we got to the *pharmacie* they asked Carole for our medical ID so we could get everything for free. Carole's French is so good that they thought we were French and had the universal ID for medical care. They informed us, sadly, that we had to pay when Carole indicated that we were not French citizens. The painkillers cost 18 euros; the crutches cost about 40 euros. This particular *pharmacie* did not have the *orthomarché*; we would need to get it at a bigger *pharmacie* in Carpentras, a slightly larger city 16 kilometers (10 miles) north of Cavaillon, the next day when it would be open again.

Our trip to Carpentras the next day might have been the most stressful part of this ordeal. Carpentras is an old city with winding streets and very few street signs. Since French directions are interesting and colorful but not necessarily precise, we wandered around for a while until we accidentally found the *pharmacie* that had the *orthomarché*. It cost another 30 or 40 euros—the exchange rate at this time was about a euro for a dollar.

When we returned to the States after this trip, I did go see my family physician and took the French X-ray with me. After I had waited 30 minutes past my appointment time, the doctor came in, looked at the ankle for 30 seconds, looked at the X-ray, and said goodbye. That minute and a half cost my insurance company and me $100. Six months later the French hammer fell. We got a bill from the Cavaillon Hôpital for 100 euros, which included the emergency room, the doctor's fee, and the X-ray.

LANDSCAPE MISCELLANEA

———————

May your trails be crooked, winding, lonesome,
dangerous, leading to the most amazing view. May
your mountains rise into and above the clouds.
—EDWARD ABBEY

Everything ends this way in France—everything.
Weddings, christenings, duels, burials, swindlings, diplomatic
affairs—everything is a pretext for a good dinner.
—JEAN ANOUILH

WHEN A BOOK SUCH AS THIS IS WRITTEN, the big picture of a place usually stands out and is the dominant theme. That is invariably appropriate. But sometimes this larger view begs for some little-picture scenarios that give the text a more human inclination. It is not uncommon for these small, idiosyncratic miscellanea to help bring into sharper focus the intricacies of landscape. What follows is a collection of these small pictures that are not necessarily directly related to each other but that, taken together, give the book a fuller texture of the two valleys.

I have talked about luscious wines and succulent fruit and exquisite dinners. But there may be no more evocative experience of the two valleys than the smell of new-mown hay in the fields at dusk. If you closed your eyes, you could not tell if you were in Provence or the North Fork Valley. That sweet, earthy odor is part of the beauty of these places. But of course that beauty is a counterpoint to what happens when the sheep are moved down to nibble on the stubble left after the hay is cut. The calm bleating of the flock is accompanied by the annoying buzzing of thousands *des mouches*—the flies that are everywhere and that get into everything, including the exquisite dinner or charming glass of wine you are trying to enjoy while dining on the lovely outdoor terrace. You take the one with the other—a sort of Coulon or North Fork yin and yang.

As we have seen, eating outside on a western Colorado or Provençal summer day is de rigueur. Part of the culture for a hundred years or more in Provence and more recently in western Colorado is to eat as many meals on a patio or terrace or balcony or deck as possible. Carole and I visited one of the beautiful Luberon villages in early July one trip. The weather was not cooperating with the Provençal tourist industry, and drizzly rains fell for days. But no one, I mean no one, would have thought of eating inside. People were huddled under awnings and umbrellas, and the wait staff was wearing sweatsuits and rain gear, but eat and drink outside we did. I think it must have been a determined stance not to miss a single solar ray if one happened to peek through the clouds, even for an instant. In the summer one eats and drinks outside and worships the sun or even the minutest possibility of sun, period.

A favorite restaurant in Paonia sits on a large plot of land in town. Customers vie for a few parking spots along the ragged street edge and enter the property through an old, non-functioning gate that is always left open. The first thing we notice in the outdoor courtyard is the gigantic cottonwood tree that dominates the space. The low-hanging branches seem to invite us to hang a rope and a tire swing—my guess is that neither the guests nor the owners would view this as inappropriate. The food here is prepared with skill and kindness, and an evening spent at one of the tables either inside or out goes by in slow, measured peacefulness.

Other, more serious aspects of the human stories in the two valleys deserve mention. The French record concerning Jews during World War II is not exemplary. Very few places we visited, and we visited a lot of them all over France, gave any public acknowledgment of the plight of the Jews (or any other designated expendable or feared groups, such as communists or gays or union officials) during that era. We found one refreshing, if depressing, instance where the loss of lives during the "deportations" was publicly remembered. The small town of Goult has the obligatory war memorial where the soldiers killed during the two world wars are remembered. But this particular obelisk also has a list of the *"Morts en Déportation"*—literally, the dead from deportation. I am sure other towns have similar monuments, but this is the only one we found during our travels. This civic honesty made me like this village just a bit more than others.

Lest we Americans feel somehow superior, the North Fork area was part of a scene that was decisive in the US–Native

GUERRE 1939-1940

	CLASSE
BRIEUNE ALBERT	1922
BOURDON ADRIEN	1933
BARADEL JULIEN	1936
CHEVALIER JEAN	1931
ESPOURTEAU PAUL	1930
GUITON LOUIS	1922
HARDY BERTIN RESISTANT	1918-1944
MORTS EN DEPORTATION	
CHAIX ROBERT	1921-1943
SWERMAN JOSEPH	1924-1944
ISAAC SAMIS	1908-1944

An unusual obelisk in the village square of Goult that included deported Jews in the list of those who died in World War II.

American wars during the late nineteenth century and the removal of the original occupants to reservations. In September 1879, Utes under the control of the US Indian Agency rebelled near Meeker, about eighty miles north of Hotchkiss. The Utes were being forced against their will to become Christians and farmers. Finally, the tension exploded into an uprising that left nearly two dozen whites dead and the band of Utes who had killed them fleeing south, toward the Grand Mesa—the northern backdrop of the North Fork. This band was captured, and a "Utes Must Go" campaign was started in Colorado. The campaign ended in the total "evacuation" of the Utes from most of the state, making it easier for ranchers, farmers, and especially miners to take over former Ute lands. Those Utes who were still left in the state were moved to reservations along Colorado's southern border, where they continue to occupy the land. I have not found any monuments to this or similar events in my travels around the North Fork either.

Even today this part of the West is not immune to a bit of xenophobia. In a 2008 essay in the *High Country News*, Paolo Bacigalupi lamented a lunchtime visit to a local café after President Barack Obama's election. He noted the pointed, not funny racist "jokes" and stereotypical comments made about President-elect Obama and his ethnic background. This is probably not an isolated case and is a very distasteful aspect of this otherwise stunning land.

On a much more pleasant and nostalgic topic that relates to Provence, one of my wife's fondest memories of her French grandfather is of him painting exquisite watercolor landscapes

A crèche of santons *collected from a typical atelier in Provence. Historically, these small figures took the place of larger village* crèches *after the French Revolution.*

of street and port scenes in and near his beloved Marseille. He would take sketch pads and paints on his frequent walks and create evocative renditions of life in southern France. He would also remember his grandchildren in America with hand-painted cards for their birthdays and Christmas. For the Yule season he often painted small crèche scenes on the cards using local *santons* as models. The making of clay *santons* began in the late eighteenth century when the French Revolution spurred a ban on public religious displays such as large crèches in front of churches. People made and bought the small handcrafted plaster pieces to set up diorama stables in the privacy of their own homes. The ban on outside displays has died away, but the Provençal tradi-

172

tion of *les santons* has grown and expanded to include characters from everyday life, comic figures, and an even greater variety of religious figures. We visited several *santon ateliers* (workshops) and often found that the shops had special importance to French culture, as evidenced by the presidential decrees framed and hung on the atelier walls. Seeing these framed decrees, one could reasonably assume that the French president has visited each of these small, usually one-person, shops.

Visiting these *santon ateliers* and seeing the small, out-of-the-way workshops these dedicated artisans inhabit is fun and always interesting. That is, of course, if they are open. As discussed in previous chapters, store hours and even times for the government post office are a bit limited and chaotic, as many of these establishments are open only ten to twenty hours a week. The scheduling of these times can be quite unusual. A week's business hours might look something like this:

Ouvert (open)	10h–12h	*lundi* (Monday)
	3h–6h	*mardi* (Tuesday)
Fermé (closed)		*mercredi* (Wednesday)
	10h–16h	*jeudi* (Thursday)
	13h–16h	*vendredi* (Friday)
Fermé		*samedi* (Saturday)
Fermé		*dimanche* (Sunday)

Or, you might check these hours assiduously and arrive precisely at 11 a.m. on *jeudi* and the place will be locked up like a jewelry store at midnight. There might be a hand-drawn sign saying *"exceptionnel fermeture"*—an exceptional or special closing. I have not made a rigorous, analytical study of these exceptional closures, but, anecdotally, they happen with a regularity that is suspicious, especially in small villages. My guess is that the cafés might be especially busy during these times. This is another of the charms of life in small French villages—the open-twenty-four-hours-a-day United States could take a hint.

The art scene in the North Fork is very young by Coulon standards, but it is vibrant and energized. For example, in Paonia the Blue Sage Center for the Arts started in a rented space in 1994. By 2000 it had raised enough money to buy an entire building in the center of town. The arts in Paonia were supported at such a level that this space proved too small, and in 2004 the center purchased the building next door so it could expand. This is an amazingly rapid development for an art center anywhere, much less in a small coal-mining town in western Colorado. Hotchkiss renovated and opened its own art center in an old creamery in 2006, and the Creamery Arts Center now caters to the increasing number of local and regional artists and artisans drawn to the valley.

So far in this book, I have said little about road systems and driving comparisons between the two countries. By far the majority of roads in the North Fork Valley area are rough gravel. Certainly the main highways, some county roads, and most town streets are paved, but the maze of roads in the valley and especially in the foothills and surrounding mountains is gravel at best. Some are good and some are . . . well, there are places in the United States where four-wheel-drive vehicles are status symbols; not here. Many roads are impassible by other than four-wheel-drive, and many of the roads at higher elevations are completely impassible in winter.

In the Vaucluse almost all of the roads are paved, no matter where they go. The only exceptions are narrow tracks into

the *garrigue* that might be used by the charcoalers or hunt-
ers. But a good description of even the "larger" departmental
("D") roads—the equivalent of a state highway system in the
United States—is that they are wide enough for one-and-a-half
vehicles, and the edges are often defined by rocks or ditches.
Almost all "D" roads have worn grass shoulders where cars have
to go to avoid head-on collisions with oncoming traffic. Even
the national "N" roads—the equivalent of the US highway system—are often narrow, especially if they traverse mountainous
terrain. Another description for many roads in the Coulon is "a
rock wall on one side and a wall of rock on the other." This could
be the modern version of Homer's Scylla and Charybdis. If two
cars meet head-on along one of these roads, one must pull over
if there is a convenient pull-off or back up until the road widens.
At some of the speeds at which French drivers go, it amazes me
that there are not many more traffic fatalities in the country. The
shops that sell brake linings must do a great business.

Cars are almost as much a part of French culture as they are
for Americans, and, as in the United States, they often cause
as many or more problems than they solve. One July morning
we woke up in our room at Le Mas du Loriot in Joucas and
went out into the already warm air, probably 80° F at 8:00 a.m.
The newspaper carried a Reuters story that was the headline for
the day, "Ozone Alert for Southern France." This meant that
people were supposed to cut down on their driving that day—a
warning that, as you can imagine, evoked a huge sigh of bore-
dom from most French drivers, who simply stepped on the gas
to get to their destinations faster, thus, in a way, driving less.

After all, if you spend less time in your car, it means you are driving less, *non*?

Ozone is a serious air quality problem, and there were several ozone alerts that summer. It had been very hot the entire time we were in Provence, and ozone is a product of the combustion of hydrocarbons (gasoline), intense sunlight, and high air temperatures. We were in the equivalent of an ozone perfect storm. Ozone (O_3) is a dangerous pollutant, especially for those with serious lung diseases. Ozone is an extreme oxidizer, which means it reacts very readily and much more robustly with other molecules than does regular molecular oxygen (O_2). It is akin to the difference between burning something with a butane lighter and vaporizing it with an acetylene torch. This is not good for the delicate tissue of the lungs.

I would be remiss if I did not include an episode in Colorado when a friend introduced me to someone (I will call him Mr. W) who had just moved from the Front Range of Colorado, where I live, to the North Fork Valley. My friend said she thought Mr. W would be excited to learn about this upcoming book. Mr. W was polite and congenial in the social situation in which we found ourselves. We made the usual noises about getting together sometime, and I planned to do just that. However, I found out a few days later that he was very determined to keep any good publicity about the North Fork Valley from reaching anyone else. He had moved to the valley just in time and did not want a bunch of other people moving in to ruin the area.

Mr. W's attitude might be justified; it is at least understandable. The problem arises when people move to a very nice place, such as the North Fork or the Vaucluse. If they come with a commitment to make the place home and to work in the community to make it even better than it is, there is no problem. But if they come to "get" only the cachet of the place without a commitment to the community and its best interests, the protectiveness seen in Mr. W's attitude surfaces.

Needless to say, I never have interviewed Mr. W for this book. In a way I am being untrue to myself when I write good things about anywhere in Colorado—I also feel that too many people have moved here without a commitment to Colorado, and what I came for decades ago is being threatened. So, Mr. W, I am empathetic. Maybe my meager efforts to "sell" this place called the North Fork will come to naught, or maybe I have been so good at selling the Vaucluse that everyone will move there instead.

Both valleys contain a collection of landscapes that are bucolic and fecund. They exhibit a long line of human occupation—sometimes very intensely settled, sometimes not. The lands have, to a great extent, been tamed by human presence and made to provide for human needs and wants. But they are also places of brutal natural beauty, with severe, harsh aspects that add a particularly piquant spiciness to otherwise domesticated places. Western Colorado has rugged foothills and tall volcanic mountains that can be numbingly cold or hot and sere, depending on the season or even the time of day. *La garrigue* and the mountains of the Vaucluse can be just as harsh and seemingly impenetrable. If the long, hot, dry summers do not get your attention, a 100-mile-an-hour mistral wind in January certainly will.

The real peoples of the lands in these two places are also rugged, hard, spare, and intense at times and gregarious, kind, generous, and fun at others—often these characteristics coexist. Neither place is Nirvana or paradise. But they may be as close as one would wish for in a life that seeks challenges, opportunities, wonder, and zest—and maybe a little loneliness and danger thrown in if Edward Abbey has any say.

THE FINISH / *C'EST FINI*

*The voyage of discovery is not in seeking new
landscapes but in having new eyes.* (redux)

—MARCEL PROUST

I REPEAT PROUST'S QUOTE because it was an appropriate
start to the book, but it is an even more suitable closing.
The basic characteristics of landscapes have been out-
lined innumerable times and in various ways by geographers, anthropologists, and landscape architects. But basically, a
landscape has three core elements, including the foundation—
the geomorphology and geology of the earth; the biota (mostly
vegetation) supported by that foundation; and the human influences, impacts, and alterations of these two physical factors. This
is pretty prosaic stuff. The poetry of landscape evolves through
the myriad ways we humans sense it—metaphorically, by having

new eyes looking upon the land where we sense the place and put it into our own personal context of understanding.

These "new eyes" can perceive and interpret the land in good ways or bad. It would be the epitome of Pollyanna-ness to assume that new discoveries will always be sanguine. In this light, in *Out of Eden* Alan Burdick observes that the world in Darwinian terms of the individual is nasty, brutish, often deadly, and short. Yet the world of the aggregate is remarkably "quiescent, functional, persistent, and durable." The two worlds of the Coulon and the North Fork are just such landscapes—the brutal, severe one at the individual level and the much more sanguine, holistic aggregate. One can see the combination and interaction of all the living things and the environmental base upon which they depend and relate. We can see the bucolic whole; unless we get close to the ground with our senses, we probably miss the intense nature of life on the edge in some remarkably harsh places. But the juxtaposition of the harsh, wild uplands of the piñon-juniper woodland and *la garrigue* and the fruited, fecund farmland of the valleys is one of the things most endearing to me as a viewer of the landscape and most strikingly similar between these two places. It is also a testament to the grit and resistance of human occupation of both of these lands and landscapes.

The individual and aggregate dichotomy Burdick describes is not only reserved for the natural world of the valleys. Many examples of human dichotomies exist in both places as well.

For instance, at a nice inn in the Luberon, we noticed an inn employee who always seemed to be at work when we went to breakfast early in the morning. He was also working during the heat of the day, around noon, and he was still doing odd jobs at the inn when we finished dinner late in the evening. By our count this was at least a twelve-hour day every day we stayed at the inn. It turned out that he was an immigrant from some Eastern European country—we could never determine which one exactly. He spoke no English and, as far as we could detect, very little French. Apparently, the easy movement of people within the European Community (EU) fostered the movement of these work-related migrants to the wealthier EU countries such as France and others. But this scenario could easily have been straight from the North Fork Valley or nearly any US agricultural area where migrants from Mexico and Central America are just as common, if not more so. They, too, often work long hours, in the heat of the day, and for depressingly low wages. Both Europe and the United States are struggling to come up with good solutions to the immigrants' plight and to protect jobs; in the meantime, people from less fortunate countries try to make a legal, and sometimes illegal, life for themselves and their families.

The landscapes of the two valleys have one critical thing in common, the reason they are near-clones of each other in the broader dimension: they are both human-scale places. The towns and villages are all easily walkable, the fields are small and individually tended, the trails are suitable for walking and biking, the food comes from local farmers as much as possible, and the wine is

The harvest of chardonnay grapes is under way in mid-October. With Yvon's joie de vivre, he gets many friends to help with the harvest.

a personal statement from the vintner, not a corporate artifact. The humanness of the land and its citizens makes the discoveries more intimate and genuine. Being diminutive, the landscapes can be sensed as a whole of the place. With a new eye and a new willingness to see, these two spots on the earth's surface are ours for the taking, even if others claim ownership.

It was a cool, sunny October weekend in the North Fork, that kind of quintessential fall Colorado day when you just cannot stay inside. The temperature the night before had been cold but not yet down to freezing. Yvon made a quick decision, or at least as quickly as Yvon makes any decision, to harvest the acre of chardonnay grapes he leases from a friend. It had been a hard year, with early frosts and not much rain, and the crop of grapes would only be about half what it had been the year before. But a small cadre of shears-wielding pickers, all friends of Yvon, took to the small vineyard with gusto. No mechanical picking here as we meticulously hunted for the pastel green grape bunches well hidden among the darker green foliage of the vines—one person to each side of a row of vines, each checking on the other so no bunch was missed. As we picked we had to scour each bunch for bad grapes and other debris Yvon insisted needed to be discarded. Each bunch was dropped into a very unromantic plastic bucket. Once a bucket was full, we dropped the precious cargo into a large plastic bin that could contain up to 800 pounds of grapes. With the enthusiastic crew and a small forklift to haul the bins, we were done with the acre in just a few hours. We worked up a thirst that had to be quenched by water, since Yvon's un-oaky chardonnay would not be ready for some time. Carole and I drove home stone-sober after a tiring, productive, fun time in the field.

At least for me, there is something special about vineyards and wineries. I cannot say exactly what it is, but I always get a feeling of land being well used, to hearken back to Wendell Berry. The growing of grapes and the "vinting" of wine cannot be done haphazardly or with a lack of diligence—it simply will not work. The lands of the vineyards in the North Fork and the Coulon are like vineyards around the world—worked intensively, with respect for the natural elements, and often just to see how well a wine can be made from the particular *terroir*

A bin of about 800 pounds of chardonnay grapes fresh from the vines on Rogers Mesa and ready to be taken to the winery.

one has available and uses. In these two hidden valleys, the wine tastes just fine.

This brings me to the end of this narrative. My initial "aha" moment gazing out the upstairs window at the slanting morning light falling on the lands of the North Fork Valley has only strengthened during intense travels in both landscapes. I know each of them better now. I have become much more of a viewer of the small likenesses and differences while still looking on in appreciation for the bigger landscapes. The smells of cultivated earth, the crackle of sere vegetation underfoot, the tastes of the new harvest, and the colors of the light as it changes with the hour and the season all make me feel a part of both the Coulon and the North Fork and make these places a part of me. These two rural regions on two different continents are certainly not clones of one another, nor would I want them to be. They are just similar enough that when I think of one, I think of the other too. But the similarities merely give me a starting point to look more closely and try to ferret out the small and meaningful differences as well. This is the meaning of geographic exploration and the reason I keep looking.

BIBLIOGRAPHY

Athanassoglou-Kallmyer, Nina. *Cézanne and Provence: The Painter in His Culture*. Chicago: University of Chicago Press, 2003, 323 pp.

Athearn, Robert G. *The Coloradans*. Albuquerque: University of New Mexico Press, 1976, 430 pp.

Balme, Christine (ed.). *Découverte Géologique du Luberon: Guide et Carte Géologique à 1/100000*. Orléans: Éditions Bureau de Recherche Géologiques et Minières, 1998, 180 pp., map.

Berry, Wendell. *Another Turn of the Crank*. Washington, DC: Counterpoint, 1995, 109 pp.

Bone, Eugenia. *At Mesa's Edge*. Boston: Houghton Mifflin, 2004, 330 pp.

Burdick, Alan. *Out of Eden*. New York: Farrar, Straus and Giroux, 2005, 324 pp.

Collison, Linda, and Bob Russell. *Rocky Mountain Wineries*. Boulder: Pruett, 1994, 165 pp.

Crouzet, Annie, and Jean-Pierre Cassely. *Provence*. Paris: Hachette Guide Evasion en France, 2008, 389 pp.

Durrell, Lawrence. *Provence*. New York: Arcade, 1990, 200 pp.

Guende, Georges, Max Gallardo, and Herve Magnin. *Parc Naturel Regional du Luberon: Secteurs de Valeur Biologique Majeure*. Apt: Parc Naturel du Luberon, 1999, 118 pp.

Guide Gallimard. *Parc Naturel Régional Luberon*. Paris: Gallimard Encyclopédies du Voyage, 2003, 192 pp.

———. *Vaucluse*. Paris: Gallimard Encyclopédies du Voyage, 2007, 372 pp.

Joseph, Robert. *French Wines: An Essential Guide to the Wines and Wine-Growing Regions of France*. New York: DK Publishing, 1999, 240 pp.

Jouval, Évelyne. *Luberon: Traces de Mémoire*. Marseille: Transbordeurs, 2004, 295 pp.

Least Heat-Moon, William. *Roads to Quoz*. New York: Little, Brown, 2008, 581 pp.

MacNeil, Karen. *The Wine Bible*. New York: Workman, 2001, 910 pp.

Mayle, Peter. *Provence A–Z*. New York: Alfred A. Knopf, 2006, 286 pp.

Meyer, Laurent. *Randonnées en Provence: Luberon*, rev. ed. Nice: Randoguides Séquoïa, 2005, 95 pp.

National Geographic Society. Trails Illustrated Map: Grand Mesa, rev. ed. Washington, DC: National Geographic Society, 2006.

Niles, Bo. *A Window on Provence*. New York: Penguin Books, 1990, 191 pp.

Roger, Frederique, and Fabrice Milochau. *Guide des Merveilles de la Nature, Provence*. Paris: Arthaud, 2004, 285 pp.

Rushford, Keith. *Reconnaître les Arbres sans Peine*. Paris: Editions Nathan, 2006, 288 pp.

Saltarelli, Jean-Pierre. *Les Côtes du Ventoux*. Avignon: Editions A. Barthelemy, 2000, 207 pp.

Smith, Alta, and Brad Smith. *The Guide to Colorado Wineries*. Golden, CO: Fulcrum, 2002, 91 pp.

Stegner, Wallace. *Beyond the Hundredth Meridian*. New York: Penguin Books, 1954, 438 pp.

Upson, Tim, and Susyn Andrews. *The Genus* Lavandula. Portland, OR: Timber, 2004, 442 pp.

Wilson, James E. *Terroir: The Role of Geology, Climate, and Culture in the Making of French Wines*. Berkeley: University of California Press, 1998, 336 pp.

Wyckoff, William. *Creating Colorado: The Making of a Western Landscape*. New Haven: Yale University Press, 1999, 336 pp.

Wylie, Laurence. *Village in the Vaucluse*, 3rd ed. Cambridge, MA: Harvard University Press, 1974, 390 pp.